Fluid Mechanics

SECOND EDITION

The multidisciplinary field of fluid mechanics is one of the most actively developing fields of physics, mathematics and engineering. This textbook, fully revised and enlarged for the Second Edition, presents the minimum of what every physicist, engineer and mathematician needs to know about hydrodynamics. It includes new illustrations throughout, using examples from everyday life: from hydraulic jumps in a kitchen sink to Kelvin–Helmholtz instabilities in clouds and geophysical and astrophysical phenomena, providing readers with a better understanding of the world around them. Aimed at undergraduate and graduate students as well as researchers, the book assumes no prior knowledge of the subject and only a basic understanding of vector calculus and analysis. It contains 41 original problems with very detailed solutions, progressing from dimensional estimates and intuitive arguments to detailed computations to help readers understand fluid mechanics.

GREGORY FALKOVICH is Professor in the Department of Physics of Complex Systems at Weizmann Institute of Science in Rehovot, Israel. He has researched in plasma, condensed matter, fluid mechanics, statistical and mathematical physics and cloud physics and meteorology, and has won several awards for his work.

Fluid Mechanics

SECOND EDITION

GREGORY FALKOVICH
Weizmann Institute of Science, Rehovot, Israel

CAMBRIDGE
UNIVERSITY PRESS

CAMBRIDGE
UNIVERSITY PRESS

University Printing House, Cambridge CB2 8BS, United Kingdom

One Liberty Plaza, 20th Floor, New York, NY 10006, USA

477 Williamstown Road, Port Melbourne, VIC 3207, Australia

314–321, 3rd Floor, Plot 3, Splendor Forum, Jasola District Centre,
New Delhi – 110025, India

79 Anson Road, #06–04/06, Singapore 079906

Cambridge University Press is part of the University of Cambridge.

It furthers the University's mission by disseminating knowledge in the pursuit of
education, learning, and research at the highest international levels of excellence.

www.cambridge.org
Information on this title: www.cambridge.org/9781107129566
DOI: 10.1017/9781316416600

© Gregory Falkovich 2018

First published 2018

Printed in the United Kingdom by TJ International Ltd. Padstow Cornwall

A catalogue record for this publication is available from the British Library.

ISBN 978-1-107-12956-6 Hardback

Contents

Preface to the Second Edition

Why study fluid mechanics? The primary reason is not even technical, it is cultural: A natural scientist is defined as one who looks around and understands at least part of the material world. One of the goals of this book is to let you understand how the wind blows and the water flows so that while flying or swimming you may appreciate what is actually going on. The secondary reason is to do with applications: Whether you are to engage with astrophysics or biophysics theory or build an apparatus for condensed matter research, you need the ability to make correct fluid-mechanics estimates; some of the art of doing this will be taught in the book. Yet another reason is conceptual: Mechanics is the basis of the whole of physics and engineering in terms of intuition and mathematical methods. Concepts introduced in the mechanics of particles were subsequently applied to optics, electromagnetism, quantum mechanics, etc.; here you will see the ideas and methods developed for the mechanics of fluids, which are used to analyze other systems with many degrees of freedom in statistical physics and quantum field theory. And last but not least: at present, fluid mechanics is one of the most actively developing fields of physics, mathematics and engineering, so you may wish to participate in this exciting development.

In the time of ever-increasing specialization, the universal language of fluid mechanics is one of the few remaining means of communications between people from very distant fields and disciplines. Using this simple and intuitive language, biophysicists and astrophysicists can tell each other about some of their recent achievements; meteorologists can comprehend what happens inside a collider while high-energy physicist can understand the cloud above his head; and physicists and mathematicians can appreciate and help to solve the problems that engineers face. My personal motivation in writing this book was to bring this language up-to-date and to express fluid mechanics using modern notions (like symmetry breaking and renormalization), most of which actually originated in the discipline.

Even for physicists who are not using fluid mechanics in their work, taking a one-semester course on the subject would be well worth the effort. This is one such course. It presumes no prior acquaintance with the subject and requires only basic knowledge of vector calculus and analysis. On the other hand, mathematicians and engineers may find in this book several new insights presented from a physicist's perspective. In choosing from the enormous wealth of material produced by the last four centuries of ever-accelerating research, preference was given to the ideas and concepts that teach lessons whose importance transcends the confines of one specific subject as they prove useful time and again across the whole spectrum of modern physics. To much delight, it turned out to be possible to weave the subjects into a single coherent narrative so that the book is a novel rather than a collection of short stories.

We approach every subject as physicists: start from qualitative considerations (dimensional reasoning, symmetries and conservation laws), then use back-of-the-envelope estimates and crown it with concise yet consistent derivations. Fluid mechanics is an essentially experimental science, as is any branch of physics and engineering. Experimental data guide us at each step, which is often far from trivial; for example, energy is not conserved even in the frictionless limit and other symmetries can be unexpectedly broken, which makes a profound impact on estimates and derivations.

Lecturers and students using the book for a course will find out that its 13 sections comfortably fit into 13 lectures plus, if needed, problem-solving sessions. Sections 2.3 and 3.1 each contain two extra subsections that can be treated in a problem-solving session (specifically, Sections 2.3.4, 2.3.5, 3.1.2 and 3.1.5, but the choice may be different). For second-year students, one can use a shorter version, excluding Sections 1.3.4, 3.2–3.4 and the small-font parts in Sections 2.2.1, 2.2.3 and 2.3.4. The lectures are supposed to be self-contained so that no references are included in the text. The epilogue and endnotes provide guidance for further reading; the references are collected in the reference list at the end. Those using the book for self-study will find out that in about two intense weeks one is able to master the basic elements of fluid mechanics. Those reading for amusement can disregard the endnotes, skip all the derivations and half of the resulting formulae and still be able to learn a lot about fluids and a bit about the world around us, helped by numerous pictures.

In many years of teaching this course at the Weizmann Institute, I have benefited from the generations of brilliant students who taught me never to stop looking for simpler explanations and deeper links between branches of physics and fields of science. Different versions of the course were taught in Lyon, Moscow, Stockholm and Stony Brook. I was also lucky to learn from many people: V. Arnold, E. Balkovsky, E. Bodenschatz, G. Boffetta, D. Budker, A. Celani, M. Chertkov, B. Chirikov, G. Eyink, U. Frisch, K. Gawedzki, V. Geshkenbein, M. Isichenko, L. Kadanoff, K. Khanin, B. Khesin, D. Khmelnitskii, I. Kolokolov, G. Kotkin, R. Kraichnan, E. Kuznetsov, A. Larkin, V. Lebedev, L. Levitov, B. Lugovtsov, S. Lukaschuk, V. L'vov, K. Moffatt, A. Newell, A. Polyakov, I. Procaccia, A. Pumir, A. Rubenchik, D. Ryutov, V. Serbo, A. Shafarenko, M. Shats, B. Shraiman, A. Shytov, E. Siggia, Ya. Sinai, M. Spektor, K. Sreenivasan, V. Steinberg, K. Turitsyn, S. Turitsyn, G. Vekshtein, M. Vergassola, P. Wiegmann, V. Zakharov, A. Zamolodchikov and Ya. Zeldovich. Special thanks to Itzhak Fouxon, Marija Vucelja and Anna Frishman, who were instructors in problem-solving sessions and helped with writing solutions for some of the exercises. I am grateful to the readers of the first English edition and the Russian edition for pointing out misprints, errors and unclear places, which I did my best to correct. Remaining errors, both of omission and of commission, are my responsibility alone. This book is dedicated to my family.

Prologue

The water's language was a wondrous one, some narrative
on a recurrent subject ...

(A. Tarkovsky, translated by A. Shafarenko)

There are two protagonists in this story: inertia and friction. One meets them first in the mechanics of particles and solids where their interplay is not very complicated: Inertia tries to keep the motion while friction tries to stop it. Going from a finite to an infinite number of degrees of freedom is always a game-changer. We will see in this book how an infinitesimal viscous friction makes fluid motion infinitely more complicated than inertia alone ever could. Without friction, most incompressible flows would stay potential, i.e. essentially trivial. At solid surfaces, friction produces vorticity, which is carried away by inertia and changes the flow in the bulk. Instabilities then bring about turbulence, and statistics emerges from dynamics. Vorticity penetrating the bulk makes life interesting in ideal fluids though in a way different from superfluids and superconductors.

On the other hand, compressibility makes even potential flows non-trivial as it allows inertia to develop a finite-time singularity (shock), which friction manages to stop. It is only in a wave motion that inertia is able to have an interesting life in the absence of friction, when it is instead partnered with medium anisotropy or inhomogeneity, which cause the dispersion of waves. The soliton is a happy child of that partnership. Yet even there, a modulational instability can bring a finite-time singularity in the form of self-focusing or collapse. At the end, I discuss how inertia, friction and dispersion may act together.

On a formal level, inertia of a continuous medium is described by a non-linear term in the equation of motion. Friction and dispersion are described by linear terms, which, however, have the highest spatial derivatives so that the limit of zero friction and zero dispersion is singular. Friction is not only singular but also a symmetry-breaking perturbation, which leads to an anomaly when the effect of symmetry breaking remains finite even in the limit of vanishing viscosity.

The first chapter introduces basic notions and describes stationary flows, inviscid and viscous. Time starts to run in the second chapter, which discusses instabilities, turbulence and sound. The third chapter is devoted to dispersive waves. It progresses from linear to nonlinear waves, solitons, collapses and wave turbulence. The epilogue gives a guide to further reading and briefly describes present-day activities in fluid mechanics. Detailed solutions of the exercises are given.

1

Basic notions and steady flows

In this chapter, we define the subject, derive the equations of motion and describe their fundamental symmetries. We start from hydrostatics where all forces are normal. We then try to consider flows this way as well, neglecting friction. This allows us to understand some features of inertia, most importantly induced mass, but the overall result is a failure to describe a fluid flow past a body. We are then forced to introduce friction and learn how it interacts with inertia, producing real flows. We briefly consider an Aristotelean world where friction dominates. In an opposite limit, we discover that the world with a little friction is very much different from the world with no friction at all.

1.1 Definitions and basic equations

Here we define the notions of fluids and their continuous motion. These definitions are induced by empirically established facts rather than deduced from a set of axioms.

1.1.1 Definitions

We deal with *continuous media* where matter may be treated as homogeneous in structure down to the smallest portions. The term *fluid* embraces both liquids and gases and relates to the fact that even though any fluid may resist deformations, that resistance cannot prevent deformation from happening. This is because the resisting force vanishes with the rate of deformation. With patience, anything can be deformed. Therefore, whether one treats the matter as a fluid or a solid depends on the time available for observation. As the prophetess Deborah sang, "The mountains flowed before the Lord"

1

(Judges 5:5). The ratio of the relaxation time to the observation time is called the Deborah number.[1] The smaller the number the more fluid the material.

A fluid can be in mechanical equilibrium only if all the mutual forces between two adjacent parts are normal to the common surface. That *experimental* observation is the basis of hydrostatics. If one applies a force parallel (tangential) to the common surface then the fluid layer on one side of the surface starts sliding over the layer on the other side. Such sliding motion will lead to a friction between layers. For example, if you cease to stir tea in a glass it could come to rest only because of such tangential forces, i.e. friction. Indeed, if the mutual action between the portions on the same radius was wholly normal, i.e. radial, then the conservation of angular momentum about the rotation axis would cause the fluid to rotate forever.

Since tangential forces are absent at rest or for a uniform flow, it is natural to consider first the flows where such forces are small and can be neglected. Therefore, a natural first step out of hydrostatics into hydrodynamics is to restrict ourselves to purely normal forces, assuming small velocity gradients (whether such a step makes sense at all and how long such approximation may last remains to be seen). Moreover, the intensity of a normal force per unit area does not depend on the direction in a fluid (Pascal's law, see Exercise 1.1). We thus characterize the internal force (or stress) in a fluid by a single scalar function $p(\mathbf{r},t)$ called pressure, which is the force per unit area. From the viewpoint of the internal state of the matter, pressure is a macroscopic (thermodynamic) variable. Microscopically, we assume every portion of the fluid to be in thermal equilibrium. In this case, the internal state of the fluid is described completely by two variables, so one needs a second thermodynamical quantity. We shall usually use the density $\rho(\mathbf{r},t)$, in addition to the pressure.

What *analytic properties* of the velocity field $\mathbf{v}(\mathbf{r},t)$ do we need to presume? We suppose the velocity to be finite and a continuous function of $\mathbf{r}$. In addition, we suppose the first spatial derivatives to be everywhere finite. That makes the *motion continuous*, i.e. trajectories of the fluid particles do not cross. The equation for the distance $\delta\mathbf{r}$ between two close fluid particles is $d\delta\mathbf{r}/dt = \delta\mathbf{v}$ so, mathematically speaking, the finiteness of $\nabla\mathbf{v}$ is the Lipschitz condition for this equation to have a unique solution (a simple example of nonunique solutions for non-Lipschitz equation is $dx/dt = |x|^{1-\alpha}$ with *two* solutions, $x(t) = (\alpha t)^{1/\alpha}$ and $x(t) = 0$, starting from zero for $\alpha > 0$). For a continuous motion, any surface moving with the fluid completely separates matter on the two sides of it. We don't yet know when exactly the continuity assumption is consistent with the equations of the fluid motion. Whether velocity derivatives may turn into infinity after a finite time is a subject of active research for an incompressible viscous fluid (and a subject of a one-million-dollar Clay prize). We shall see that a compressible inviscid flow generally develops discontinuities, called shocks.

1.1.2 Equations of motion for an ideal fluid

The Euler equation. The force acting on any fluid volume is equal to the pressure integral over the surface: $-\oint p\,\mathrm{d}\mathbf{f}$. The surface area element $\mathrm{d}\mathbf{f}$ is a vector directed as outward normal:

Let us transform the surface integral into the volume one: $-\oint p\,\mathrm{d}\mathbf{f} = -\int \nabla p\,\mathrm{d}V$. The force acting on a unit volume is thus $-\nabla p$. That would be wrong, however, to assume that this force is the time derivative of the momentum $\rho\mathbf{v}$ of this volume. To write the second law of Newton, we need to single out a fixed body of fluid. An infinitesimal such body is called *fluid particle* and it always contains the same mass, which we assume unity. Then the force per unit mass, $\nabla p/\rho$, must be equal to the acceleration $\mathrm{d}\mathbf{v}/\mathrm{d}t$:

$$\frac{\mathrm{d}\mathbf{v}}{\mathrm{d}t} = -\frac{\nabla p}{\rho} \ .$$

The acceleration $\mathrm{d}\mathbf{v}/\mathrm{d}t$ is not the rate of change of the fluid velocity at a fixed point in space but the rate of change of the velocity of a given fluid particle as it moves about in space. One uses the chain rule of differentiation to express this (substantial or material) derivative in terms of quantities referring to points fixed in space. During the time $\mathrm{d}t$ the fluid particle changes its velocity by $\mathrm{d}\mathbf{v}$ (which is composed of two parts, temporal and spatial):

$$\mathrm{d}\mathbf{v} = \mathrm{d}t\frac{\partial \mathbf{v}}{\partial t} + (\mathrm{d}\mathbf{r}\cdot\nabla)\mathbf{v} = \mathrm{d}t\frac{\partial \mathbf{v}}{\partial t} + \mathrm{d}x\frac{\partial \mathbf{v}}{\partial x} + \mathrm{d}y\frac{\partial \mathbf{v}}{\partial y} + \mathrm{d}z\frac{\partial \mathbf{v}}{\partial z} \ . \tag{1.1}$$

It is the change in the fixed point plus the difference at two points $\mathrm{d}\mathbf{r}$ apart, where $\mathrm{d}\mathbf{r} = \mathbf{v}\mathrm{d}t$ is the distance moved by the fluid particle during $\mathrm{d}t$ due to inertia. Dividing (1.1) by $\mathrm{d}t$ we obtain the substantial derivative as a local derivative plus a convective derivative:

$$\frac{\mathrm{d}\mathbf{v}}{\mathrm{d}t} = \frac{\partial \mathbf{v}}{\partial t} + (\mathbf{v}\cdot\nabla)\mathbf{v} \ .$$

We see that even when the flow is steady, $\partial\mathbf{v}/\partial t = 0$, the acceleration is nonzero as long as $(\mathbf{v}\cdot\nabla)\mathbf{v}\neq 0$, that is if the velocity field changes in space along itself.

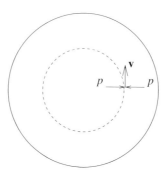

Figure 1.1 The radial pressure gradient is normal to circular surfaces and cannot change the moment of momentum of the fluid inside or outside the surface; it changes the direction of velocity **v** but not its modulus.

Any function $F(\mathbf{r}(t), t)$, like fluid temperature, varies for a moving particle in the same way, according to the chain rule of differentiation:

$$\frac{\mathrm{d}F}{\mathrm{d}t} = \frac{\partial F}{\partial t} + (\mathbf{v} \cdot \nabla)F .$$

Writing now the second law of Newton for a unit mass of a fluid, we come to the equation derived by Euler (Berlin 1757; Petersburg 1759):

$$\frac{\partial \mathbf{v}}{\partial t} + (\mathbf{v} \cdot \nabla)\mathbf{v} = -\frac{\nabla p}{\rho} . \qquad (1.2)$$

Before Euler, the acceleration of a fluid had been considered as due to the difference of the pressure exerted by the enclosing walls. Euler introduced the pressure field *inside* the fluid. For example, for the steadily rotating fluid shown in Figure 1.1, the acceleration vector $(\mathbf{v} \cdot \nabla)\mathbf{v}$ has a nonzero radial component v^2/r. The radial acceleration times the density gives the radial pressure gradient: $\mathrm{d}p/\mathrm{d}r = \rho v^2/r$.

We can also add an external body force per unit mass (for gravity $\mathbf{f} = \mathbf{g}$):

$$\frac{\partial \mathbf{v}}{\partial t} + (\mathbf{v} \cdot \nabla)\mathbf{v} = -\frac{\nabla p}{\rho} + \mathbf{f}. \qquad (1.3)$$

The term $(\mathbf{v} \cdot \nabla)\mathbf{v}$ describes inertia and makes (1.3) nonlinear.

Continuity equation. This expresses conservation of mass. If Q is the volume of a moving element then $\mathrm{d}\rho Q/\mathrm{d}t = 0$, that is

$$Q\frac{\mathrm{d}\rho}{\mathrm{d}t} + \rho\frac{\mathrm{d}Q}{\mathrm{d}t} = 0. \qquad (1.4)$$

The volume change can be expressed via $\mathbf{v}(\mathbf{r}, t)$.

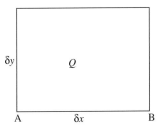

The horizontal velocity of the point B relative to the point A is $\delta x \partial v_x / \partial x$. After the time interval dt, the length of the edge AB is $\delta x (1 + dt \partial v_x / \partial x)$. Overall, after dt, one has the volume change

$$dQ = dt\, \delta x \delta y \delta z \left(\frac{\partial v_x}{\partial x} + \frac{\partial v_y}{\partial y} + \frac{\partial v_z}{\partial z} \right) = dt\, Q\, \mathrm{div}\,\mathbf{v} = dt\, \frac{dQ}{dt} \ .$$

Substituting that into (1.4) and canceling (arbitrary) Q we obtain the continuity equation

$$\frac{d\rho}{dt} + \rho\, \mathrm{div}\,\mathbf{v} = \frac{\partial\rho}{\partial t} + (\mathbf{v}\cdot\nabla)\rho + \rho\, \mathrm{div}\,\mathbf{v} = \frac{\partial\rho}{\partial t} + \mathrm{div}(\rho\mathbf{v}) = 0 \ . \qquad (1.5)$$

The last equation is almost obvious since for any *fixed volume of space* the decrease of the total mass inside, $-\int (\partial\rho / \partial t)\, dV$, is equal to the flux $\oint \rho \mathbf{v}\cdot\mathbf{df} = \int \mathrm{div}(\rho\mathbf{v}) dV$.

Entropy equation. We now have four equations (1.3, 1.5) for five quantities p, ρ, v_x, v_y, v_z, so we need one extra equation. In deriving (1.3, 1.5) we have taken no account of energy dissipation, thus neglecting internal friction (viscosity) and heat exchange. A fluid without viscosity and thermal conductivity is called *ideal*. The motion of an ideal fluid is adiabatic, that is the entropy of any fluid particle remains constant: $ds/dt = 0$, where s is the entropy per unit mass. We can turn this equation into a continuity equation for the entropy density in space

$$\frac{\partial(\rho s)}{\partial t} + \mathrm{div}(\rho s \mathbf{v}) = 0 \ . \qquad (1.6)$$

Since entropy is a function of pressure and density then (1.6) is the needed extra relation between velocity, pressure and density. Different media differ by the form of the function $s(P, \rho)$.

Boundary conditions. At the boundaries of the fluid, the continuity equation (1.5) is replaced by the *boundary conditions*:

(1) On a fixed boundary, $v_n = 0$;
(2) On a moving boundary between two immiscible fluids, $p_1 = p_2$ and $v_{n1} = v_{n2}$.

These are particular cases of the general surface condition. Let $F(\mathbf{r},t) = 0$ be the equation of the bounding surface. An absence of any fluid flow across the surface requires

$$\frac{\mathrm{d}F}{\mathrm{d}t} = \frac{\partial F}{\partial t} + (\mathbf{v} \cdot \nabla)F = 0,$$

which means, as we now know, the zero rate of F variation for a fluid particle. For a stationary boundary, $\partial F/\partial t = 0$ and $\mathbf{v} \perp \nabla F \Rightarrow v_n = 0$.

1.1.3 Hydrostatics

A necessary and sufficient condition for fluid to be in a mechanical equilibrium follows from (1.3):

$$\nabla p = \rho \mathbf{f}. \tag{1.7}$$

Not every distribution of $\rho(\mathbf{r})$ could be in equilibrium since $\rho(\mathbf{r})\mathbf{f}(\mathbf{r})$ is not necessarily a gradient. If the force is potential, $\mathbf{f} = -\nabla\phi$, then taking the *curl* of (1.7) we get

$$\nabla\rho \times \nabla\phi = 0.$$

This means that the gradients of ρ and ϕ are parallel and their level surfaces coincide in equilibrium. The best-known example is gravity with $\phi = gz$ and $\partial p/\partial z = -\rho g$. For an incompressible fluid, it gives

$$p(z) = p(0) - \rho g z.$$

For an ideal gas under a homogeneous temperature, which has $p = \rho T/m$, one gets

$$\frac{\mathrm{d}p}{\mathrm{d}z} = -\frac{pgm}{T} \quad \Rightarrow \quad p(z) = p(0)\exp(-mgz/T).$$

For air at $0\,°\mathrm{C}$, $T/mg \simeq 8$ km. The Earth's atmosphere is described by neither a linear nor an exponential law because of an inhomogeneous temperature (Figure 1.2). Assuming a linear temperature decay, $T(z) = T_0 - \alpha z$, one obtains a better approximation:

$$\frac{\mathrm{d}p}{\mathrm{d}z} = -\rho g = -\frac{pmg}{T_0 - \alpha z},$$
$$p(z) = p(0)(1 - \alpha z/T_0)^{mg/\alpha},$$

which can be used not far from the surface with $\alpha \simeq 6.5\,°\mathrm{C}\,\mathrm{km}^{-1}$.

Under gravity, density depends only on the distance from the Earth center (or locally on the vertical coordinate z) in a mechanical equilibrium. According

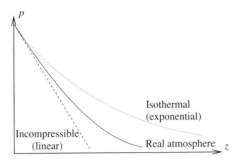

Figure 1.2 Pressure–height dependence for an incompressible fluid (broken line), isothermal gas (dotted line) and a real atmosphere (solid line).

to $dp/dz = -\rho g$, the pressure also depends only on z. Pressure and density determine temperature, which must then also be independent of the horizontal coordinates. Different temperatures at the same height, in particular nonuniform temperature of the Earth surface, necessarily produce fluid motion, which is why winds blow in the atmosphere and currents flow in the ocean. Another source of atmospheric flows is thermal convection due to a negative vertical temperature gradient. Let us derive the stability criterion for a fluid with a vertical profile $T(z)$. If a fluid element is shifted up adiabatically from z by dz, it keeps its entropy $s(z)$ but acquires the pressure $p' = p(z+dz)$ so its new density is $\rho(s, p')$. For stability, this density must exceed the density of the displaced air at the height $z + dz$, which has the same pressure but different entropy $s' = s(z + dz)$. The condition for stability of the stratification is as follows:

$$\rho(p', s) > \rho(p', s') \Rightarrow \left(\frac{\partial \rho}{\partial s} \right)_p \frac{ds}{dz} < 0 \,.$$

Entropy usually increases under expansion, $(\partial \rho / \partial s)_p < 0$, and for stability we must require $ds/dz > 0$. Entropy depends on p, T which both decay with the height. Entropy decreases with cooling yet increases when P decreases. To see which effect wins we compute:

$$\frac{ds}{dz} = \left(\frac{\partial s}{\partial T} \right)_p \frac{dT}{dz} + \left(\frac{\partial s}{\partial p} \right)_T \frac{dp}{dz} = \frac{c_p}{T} \frac{dT}{dz} + \left(\frac{\partial V}{\partial T} \right)_p \frac{g}{V} > 0. \tag{1.8}$$

Here we used specific volume $V = 1/\rho$. For an ideal gas the coefficient of the thermal expansion gives $(\partial V / \partial T)_p = V/T$ and we end up with

$$\frac{g}{c_p} > -\frac{dT}{dz} \,. \tag{1.9}$$

Indeed, stability requires that the gain in potential energy gdz must exceed the decrease in thermal energy $c_p dT$. For the Earth's atmosphere, $c_p \sim 10^3 \mathrm{J/kg^{-1}K^{-1}}$ and the convection threshold is $10°\mathrm{C\,km^{-1}}$. The average gradient is $6.5°\mathrm{C\,km^{-1}}$, that is, the entropy decreases with the height and the atmosphere is globally stable. However, local gradients vary very much depending on ground albedo, evaporation, etc., so that the atmosphere is often locally unstable with respect to thermal convection. The human body always excites convection in room-temperature air.[2]

Temperature decays with height only in the troposphere that is until about $-50°\mathrm{C}$ at 10–12 km, it is then constant up to about 35 km so that the pressure decays exponentially, eventually it grows in the stratosphere until about $0°\mathrm{C}$ at 50 km. Looking down from the plane flying above 10 km one often sees flat cloud top, particularly so-called anvil clouds, which is exactly where unstable air stratification below turns into stable above.

The convection stability argument applied to an incompressible fluid rotating with the angular velocity $\Omega(r)$ gives the Rayleigh's stability criterion, $\mathrm{d}(r^2\Omega)^2/\mathrm{d}r > 0$, which states that the angular momentum of the fluid $L = r^2|\Omega|$ must increase with the distance r from the rotation axis.[3] Indeed, if a fluid element is shifted from r to r' it keeps its angular momentum $L(r)$, so that the local pressure gradient $\mathrm{d}p/\mathrm{d}r = \rho r'\Omega^2(r')$ must overcome the centrifugal force $\rho r'(L^2 r^4/r'^4)$.

1.1.4 Isentropic motion

The simplest motion corresponds to constant s and allows for a substantial simplification of the Euler equation. Indeed, it would be convenient to represent $\nabla p/\rho$ as a gradient of some function. For this end, we need a function that depends on p, s, so that at $s = \mathrm{const.}$ its differential is expressed solely via $\mathrm{d}p$. There exists the thermodynamic potential called *enthalpy*, defined as $W = E + pV$ per unit mass (E is the internal energy of the fluid). For our purposes, it is enough to remember from thermodynamics the single relation $\mathrm{d}E = T\mathrm{d}s - p\mathrm{d}V$ so that $\mathrm{d}W = T\mathrm{d}s + V\mathrm{d}p$ (one can also show that $W = \partial(E\rho)/\partial\rho$). Since $s = \mathrm{const.}$ for an isentropic motion and $V = \rho^{-1}$ for a unit mass, $\mathrm{d}W = \mathrm{d}p/\rho$ and, without body forces one has

$$\frac{\partial \mathbf{v}}{\partial t} + (\mathbf{v} \cdot \nabla)\mathbf{v} = -\nabla W. \qquad (1.10)$$

Such a gradient form will be used extensively for obtaining conservation laws, integral relations, etc. For example, we can use the vector identity $\mathbf{A} \times (\nabla \times \mathbf{B}) = \mathbf{A} \cdot (\nabla \mathbf{B}) - (\mathbf{A} \cdot \nabla)\mathbf{B}$ to represent

$$(\mathbf{v} \cdot \nabla)\mathbf{v} = \nabla v^2/2 - \mathbf{v} \times (\nabla \times \mathbf{v}),$$

and get

$$\frac{\partial \mathbf{v}}{\partial t} = \mathbf{v} \times (\nabla \times \mathbf{v}) - \nabla(W + v^2/2). \qquad (1.11)$$

The first term on the right-hand side is perpendicular to the velocity. To project (1.11) along the velocity and get rid of this term, we define a streamline as a line whose tangent is everywhere parallel to the instantaneous velocity. The streamlines are then determined by the relations

$$\frac{dx}{v_x} = \frac{dy}{v_y} = \frac{dz}{v_z} .$$

Note that for time-dependent flows streamlines are different from particle trajectories: tangents to streamlines give velocities at a given time while tangents to trajectories give velocities at subsequent times. One records streamlines experimentally by seeding fluids with light-scattering particles; each particle produces a short trace on a short-exposure photograph, and the length and orientation of the trace indicates the magnitude and direction of the velocity. Streamlines can intersect only at a point of zero velocity called the stagnation point.

Let us now consider a steady flow, assuming $\partial \mathbf{v}/\partial t = 0$, and take the component of (1.11) along the velocity at a point:

$$\frac{\partial}{\partial l}(W + v^2/2) = 0 . \qquad (1.12)$$

We see that $W + v^2/2 = E + p/\rho + v^2/2$ is constant along any given streamline, but may be different for different streamlines (Bernoulli 1738). Bernoulli theorem, of course, is a particular case of energy conservation. The change of the total energy density is not zero along the streamline but is equal to $P_2/\rho_2 - P_1/\rho_1$ which is the work done. This is the reason W rather E enters the conservation law, as also discussed after (1.18). Alternatively, one may say that W is a potential energy of a fluid particle, see (1.41) below. In a gravity field,

$$W + gz + v^2/2 = \text{const.} \qquad (1.13)$$

Without much exaggeration, one can say that most fluid-mechanics estimates use (1.12) or (1.13). Let us consider several applications of this useful relation.

Imagine that our spaceship suffered a meteorite attack that left holes in the walls of the cabin and the tank with liquid fuel. We need to estimate how fast we lose air from the cabin and fuel from the tank. Since there is vacuum outside, we can neglect thermal exchange and consider both flows isentropic. Liquid could be treated as incompressible, its internal energy E is then constant without any external force. Bernoulli theorem then gives the limiting velocity with which such a liquid escapes from a large reservoir into vacuum:

$$v = \sqrt{2p_0/\rho} \ .$$

For water ($\rho = 10^3 \, \text{kg m}^{-3}$) at atmospheric pressure ($p_0 = 10^5 \, \text{N m}^{-2}$) one gets $v = \sqrt{200} \approx 14 \, \text{m s}^{-1}$.

For a gas, pressure drop must be accompanied by density change. The adiabatic law, $p/p_0 = (\rho/\rho_0)^\gamma$, gives the enthalpy as:

$$W = \int \frac{\mathrm{d}p}{\rho} = \frac{\gamma p}{(\gamma-1)\rho} \ .$$

The limiting velocity for the escape into vacuum can again be found from Bernoulli theorem:

$$\frac{\gamma p_0}{(\gamma-1)\rho} = \frac{v^2}{2} \quad \Rightarrow \quad v = \sqrt{\frac{2\gamma p_0}{(\gamma-1)\rho}},$$

The velocity is $\sqrt{\gamma/(\gamma-1)}$ times larger than for an incompressible fluid which corresponds to the limit $\gamma \gg 1$. The gas flows faster because the internal energy of the gas decreases as it flows, thus increasing the kinetic energy. We conclude that a meteorite-damaged spaceship loses the air from the cabin faster than the liquid fuel from the tank. We shall see later that $(\partial P/\partial\rho)_s = \gamma P/\rho$ is the sound velocity squared, c^2, so that $v = c\sqrt{2/(\gamma-1)}$. For an ideal gas with n internal degrees of freedom, $W = E + p/\rho = nT/2m + T/m$ so that $\gamma = (2+n)/n$. For bi-atomic molecules $n = 5$ (3 translations and 2 rotations) at not very high temperature, when vibrations are not excited.

Another frequent occurrence is efflux from a small orifice under the action of gravity. Supposing the external pressure to be the same at the horizontal surface and at the orifice, we apply the Bernoulli relation to the streamline which originates at the upper surface with almost zero velocity and exits with velocity $v = \sqrt{2gh}$ (Torricelli 1643). The Torricelli formula is not of much use practically to calculate the rate of discharge, which in reality is not equal to the orifice area times $\sqrt{2gh}$, the fact known to wine merchants long before physicists. Indeed, streamlines converge from all sides toward the orifice so that the jet continues to converge for a while after coming out (Figure 1.3). Moreover, the converging motion makes the pressure in the interior of the jet somewhat greater than that at the surface (as is clear from the curvature of streamlines) so that the velocity in the interior is somewhat less than $\sqrt{2gh}$. The experiment shows that contraction ceases and the jet becomes cylindrical

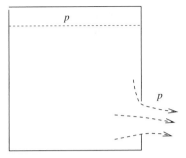

Figure 1.3 Streamlines converge coming out of the orifice.

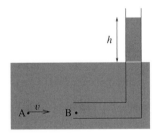

Figure 1.4 Pitot tube, which determines the velocity v at the point A by measuring the height h.

at a short distance beyond the orifice. This point is called "vena contracta" and the ratio of the jet area there to the orifice area is called the coefficient of contraction. The estimate for the discharge rate is $\sqrt{2gh}$ times the orifice area times the coefficient of contraction. For a round hole in a thin wall, the coefficient of contraction is experimentally found to be 0.62. Exercise 1.3 presents a particular case where the coefficient of contraction can be found exactly.

The Bernoulli relation is also used in different devices that measure the flow velocity. Probably, the simplest such device is the *Pitot tube* shown in Figure 1.4. It is open at both ends with the horizontal arm facing upstream. Since the liquid does not move inside the tube then the velocity is zero at the point labeled B. On the one hand, the pressure difference at two points on the same streamline can be expressed via the velocity at A: $P_B - P_A = \rho v^2/2$. On the other hand, it is expressed via the height h by which liquid rises above the surface in the vertical arm of the tube: $P_B - P_A = \rho g h$. That gives $v^2 = 2gh$.

One may wonder why the Earth atmosphere is not isentropic as remarked in the previous section. Rising water vapour condenses and releases latent heat, making the mean rate of temperature decrease lower than adiabatic.

1.2 Conservation laws and potential flows

In this section we deduce the conservation laws and their straightforward consequences from the equations of motion.

Symmetries and conservation laws. The equations of ideal hydrodynamics (1.3, 1.5, 1.6) express, respectively, the conservation laws of the momentum, mass and entropy. They are invariant with respect to space translations (which brings momentum conservation), and time translations (which brings energy conservation, described in the next subsection). The equations are time-reversible that is invariant with respect to the transformation $t \to -t$ and $\mathbf{v} \to -\mathbf{v}$ – we shall see later how the breakdown of this symmetry makes real flows so interesting. An additional symmetry is the Galilean invariance with respect to passing to a reference frame moving with the speed $\mathbf{V}$: $\mathbf{v} \to \mathbf{v} + \mathbf{V}$ and $\mathbf{r} \to \mathbf{r} - \mathbf{V}t$.[4] The equations of ideal hydrodynamics are also invariant with respect to re-scaling $\mathbf{r} \to \mathbf{r}/a$, $t \to t/b$, $\mathbf{v} \to \mathbf{v}b/a$.

Eulerian and Lagrangian descriptions. We thus encountered two alternative types of description. The equations (1.3, 1.6) use the coordinate system fixed in space, like field theories describing electromagnetism or gravity. This type of description is called Eulerian in fluid mechanics. Another approach is called Lagrangian; it is a generalization of the approach taken in particle mechanics. In this method one follows fluid particles[5] and treats their current coordinates, $\mathbf{r}(\mathbf{R},t)$, as functions of time and their initial positions $\mathbf{R} = \mathbf{r}(\mathbf{R},0)$. The substantial derivative is thus the Lagrangian derivative since it sticks to a given fluid particle, that is keeps $\mathbf{R}$ constant: $d/dt = (\partial/\partial t)_\mathrm{R}$. Conservation laws written for a unit-mass quantity $\mathscr{A}$ have a Lagrangian form:

$$\frac{d\mathscr{A}}{dt} = \frac{\partial \mathscr{A}}{\partial t} + (\mathbf{v}\nabla)\mathscr{A} = 0 \ .$$

Every Lagrangian conservation law together with mass conservation generates an Eulerian conservation law for a unit-volume quantity $\rho\mathscr{A}$:

$$\frac{\partial(\rho\mathscr{A})}{\partial t} + \mathrm{div}(\rho\mathscr{A}\mathbf{v}) = \mathscr{A}\left[\frac{\partial\rho}{\partial t} + \mathrm{div}(\rho\mathbf{v})\right] + \rho\left[\frac{\partial\mathscr{A}}{\partial t} + (\mathbf{v}\nabla)\mathscr{A}\right] = 0.$$

On the contrary, if the Eulerian conservation law has the form

$$\frac{\partial(\rho\mathscr{B})}{\partial t} + \mathrm{div}(\mathbf{F}) = 0 \qquad\qquad (1.14)$$

and the flux is not equal to the density times velocity, $\mathbf{F} \neq \rho\mathscr{B}\mathbf{v}$, then the respective Lagrangian conservation law does not exist. That means that fluid particles can exchange $\mathscr{B}$ conserving the total space integral – we shall see that the conservation laws of energy and momentum have that form.

1.2.1 Energy and momentum fluxes

Since we expect fluid particles to exchange energy and momentum then the respective fluxes must be different from "velocity times density of energy/momentum" like in (1.14). What is the difference?

The Euler equation is itself a momentum-conservation equation and must have the form of a continuity equation written for the momentum density. The momentum of the unit volume is the vector $\rho \mathbf{v}$ whose every component is conserved so it should satisfy the equation of the form

$$\frac{\partial \rho v_i}{\partial t} + \frac{\partial \Pi_{ik}}{\partial x_k} = 0 \, .$$

Let us find the momentum flux Π_{ik} – the flux of the ith component of the momentum across the surface with the normal along k. Substitute the mass continuity equation $\partial \rho / \partial t = -\partial (\rho v_k)/\partial x_k$ and the Euler equation $\partial v_i/\partial t = -v_k \partial v_i/\partial x_k - \rho^{-1}\partial p/\partial x_i$ into

$$\frac{\partial \rho v_i}{\partial t} = \rho \frac{\partial v_i}{\partial t} + v_i \frac{\partial \rho}{\partial t} = -\frac{\partial p}{\partial x_i} - \frac{\partial}{\partial x_k} \rho v_i v_k,$$

that is

$$\Pi_{ik} = p\delta_{ik} + \rho v_i v_k. \tag{1.15}$$

Plainly speaking, along $\mathbf{v}$ there is only the flux of parallel momentum $p + \rho v^2$ while perpendicular to $\mathbf{v}$ the momentum component is zero at the given point and the flux is p. For example, if we direct the x-axis along the velocity at a given point then $\Pi_{xx} = p + v^2$, $\Pi_{yy} = \Pi_{zz} = p$ and all the off-diagonal components are zero.

Let us now derive the equation that expresses the conservation law of energy. The energy density (per unit volume) in the flow is $\rho(E + v^2/2)$. For isentropic flows, one can use $\partial \rho E/\partial \rho = E + \rho \partial E/\partial \rho = E - \rho^{-1}\partial E/\partial V = E + P/\rho = W$ and calculate the time derivative

$$\frac{\partial}{\partial t}\left(\rho E + \frac{\rho v^2}{2}\right) = (W + v^2/2)\frac{\partial \rho}{\partial t} + \rho v \cdot \frac{\partial v}{\partial t} = -\mathrm{div}\left[\rho v (W + v^2/2)\right].$$

Since the right-hand side is a total derivative, the integral of the energy density over the whole space is conserved. The same Eulerian conservation law in the form of a continuity equation can be obtained in a general (non-isentropic) case as well. It is straightforward to calculate the time derivative of the kinetic energy:

$$\frac{\partial}{\partial t}\frac{\rho v^2}{2} = -\frac{v^2}{2}\operatorname{div}\rho v - v\cdot\nabla p - \rho v\cdot(v\nabla)v$$

$$= -\frac{v^2}{2}\operatorname{div}\rho v - v(\rho\nabla W - \rho T\nabla s) - \rho v\cdot\nabla v^2/2.$$

$$= -\operatorname{div}\rho vv^2/2 - v(\rho\nabla W - \rho T\nabla s). \tag{1.16}$$

For calculating $\partial(\rho E)/\partial t$ we use $\mathrm{d}E = T\mathrm{d}s - p\mathrm{d}V = T\mathrm{d}s + p\rho^{-2}\mathrm{d}\rho$ so that $\mathrm{d}(\rho E) = E\mathrm{d}\rho + \rho\mathrm{d}E = W\mathrm{d}\rho + \rho T\mathrm{d}s$ and

$$\frac{\partial(\rho E)}{\partial t} = W\frac{\partial\rho}{\partial t} + \rho T\frac{\partial s}{\partial t} = -W\operatorname{div}\rho v - \rho T v\cdot\nabla s$$

$$= -\operatorname{div}\rho vW + v(\rho\nabla W - \rho T\nabla s). \tag{1.17}$$

The right sides of (1.16, 1.17) contain divergences of the respective fluxes plus the exchange term (the last bracket) coming with opposite signs. Adding kinetic and potential energies together, one gets the exchange terms canceled:

$$\frac{\partial}{\partial t}\left(\rho E + \frac{\rho v^2}{2}\right) = -\operatorname{div}\left[\rho v(W + v^2/2)\right]. \tag{1.18}$$

As usual, the rhs is the divergence of the flux, indeed:

$$\frac{\partial}{\partial t}\int\left(\rho E + \frac{\rho v^2}{2}\right)\mathrm{d}V = -\oint\rho(W + v^2/2)\mathbf{v}\cdot\mathrm{d}\mathbf{f}.$$

As expected, the energy flux,

$$\rho\mathbf{v}(W + v^2/2) = \rho\mathbf{v}(E + v^2/2) + p\mathbf{v},$$

is not equal to the energy density times $\mathbf{v}$ but contains an extra pressure term that describes the work done by pressure forces on the fluid, similarly to the momentum flux. In other terms, any unit mass of the fluid carries an amount of energy $W + v^2/2$ rather than $E + v^2/2$. That means, in particular, that for energy there is no (Lagrangian) conservation law for unit mass $\mathrm{d}(\cdot)/\mathrm{d}t = 0$ that is valid for passively transported quantities such as entropy. This is natural because different fluid elements exchange energy by doing work.

1.2.2 Kinematics

We consider here the kinematics of a small fluid element. In particular, it will help us to appreciate the new conservation law, described in the next subsection. The relative motion near a point is determined by the velocity difference between neighboring points:

$$\delta v_i = r_j \partial v_i / \partial x_j.$$

It is convenient to analyze the tensor of the velocity derivatives by decomposing it into symmetric and antisymmetric parts: $\partial v_i / \partial x_j = S_{ij} + A_{ij}$. The symmetric tensor $S_{ij} = (\partial v_i / \partial x_j + \partial v_j / \partial x_i)/2$ is called strain. The vector initially parallel to the axis j turns toward the axis i with the angular speed $\partial v_i / \partial x_j$, so that $2S_{ij}$ is the rate of variation of the angle between two initially mutually perpendicular small vectors along i and j axes. In other words, $2S_{ij}$ is the rate with which rectangle deforms into parallelogram. Of course, we can always transform a symmetric tensor into a diagonal form by an orthogonal transformation (i.e. by the rotation of the axes). The diagonal components are the rates of stretching in different directions. Indeed, the equation for the distance between two points along a principal direction has a form: $\dot{r}_i = \delta v_i = r_i S_{ii}$ (no summation over i). The solution is as follows:

$$r_i(t) = r_i(0) \exp\left[\int_0^t S_{ii}(t') \, dt'\right].$$

For a permanent strain, the growth or decay is exponential in time. One recognizes that a purely straining motion converts a spherical material element into an ellipsoid, where the principal diameters grow (or decay) in time and do not rotate. Indeed, consider a circle of radius R at $t = 0$. The point that starts at $x_0, y_0 = \sqrt{R^2 - x_0^2}$ goes into

$$x(t) = e^{S_{11}t} x_0,$$

$$y(t) = e^{S_{22}t} y_0 = e^{S_{22}t}\sqrt{R^2 - x_0^2} = e^{S_{22}t}\sqrt{R^2 - x^2(t)e^{-2S_{11}t}},$$

$$x^2(t)e^{-2S_{11}t} + y^2(t)e^{-2S_{22}t} = R^2. \tag{1.19}$$

The equation (1.19) describes how the initial fluid circle turns into an ellipse whose eccentricity increases exponentially with the rate $|S_{11} - S_{22}|$ (Figure 1.5).

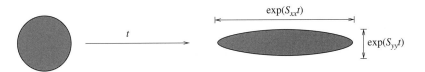

Figure 1.5 Deformation of a fluid element by a permanent strain.

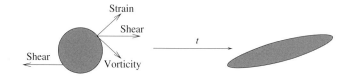

Figure 1.6 Deformation and rotation of a fluid element in a shear flow. Shearing motion is decomposed into a straining motion and rotation.

The sum of the strain diagonal components is $\mathrm{div}\,\mathbf{v} = S_{ii}$ which determines the rate of the volume change:

$$Q^{-1}\mathrm{d}Q/\mathrm{d}t = -\rho^{-1}\mathrm{d}\rho/\mathrm{d}t = \mathrm{div}\,\mathbf{v} = S_{ii}.$$

The antisymmetric part $A_{ij} = (\partial v_i/\partial x_j - \partial v_j/\partial x_i)/2$ has only three independent components so it could be represented via some vector ω: $A_{ij} = -\varepsilon_{ijk}\omega_k/2$. The coefficient $-1/2$ is introduced to simplify the relation between $\mathbf{v}$ and ω:

$$\omega = \nabla \times \mathbf{v}.$$

The vector ω is called the *vorticity* as it describes the rotation of the fluid element: $\delta\mathbf{v} = [\omega \times \mathbf{r}]/2$. It is twice the effective local angular velocity of the fluid. A plane shearing motion, such as $v_x(y)$, corresponds to strain and vorticity being equal in magnitude (Figure 1.6).

1.2.3 Kelvin's theorem

This theorem describes the conservation of velocity circulation for isentropic flows. For a rotating cylinder of a fluid, the angular momentum is proportional to the velocity circulation around the cylinder circumference. The angular momentum and circulation are both conserved when there are only normal forces, as was already mentioned at the beginning of Section 1.1.1. Let us show that this is also true for every "fluid" contour that is made of fluid particles. As fluid moves, both the velocity and the contour shape change:

$$\frac{\mathrm{d}}{\mathrm{d}t}\oint \mathbf{v}\cdot\mathrm{d}\mathbf{l} = \oint \mathbf{v}(\mathrm{d}\mathbf{l}/\mathrm{d}t) + \oint (\mathrm{d}\mathbf{v}/\mathrm{d}t)\cdot\mathrm{d}\mathbf{l} = 0.$$

The first term here disappears because it is a contour integral of the complete differential: since $\mathrm{d}\mathbf{l}/\mathrm{d}t = \delta\mathbf{v}$ then $\oint \mathbf{v}(\mathrm{d}\mathbf{l}/\mathrm{d}t) = \oint \delta(v^2/2) = 0$. In the second term we substitute the Euler equation for isentropic motion, $\mathrm{d}\mathbf{v}/\mathrm{d}t = -\nabla W$, and use Stokes' formula, which tells that the circulation of a vector around a closed contour is equal to the flux of the curl through any surface bounded by the contour: $\oint \nabla W \cdot \mathrm{d}\mathbf{l} = \int \nabla \times \nabla W\,\mathrm{d}\mathbf{f} = 0$.

Stokes' formula also tells us that $\oint \mathbf{v}\mathrm{dl} = \int \boldsymbol{\omega}\cdot\mathrm{df}$. Therefore, the conservation of the velocity circulation equals the conservation of the vorticity flux. To better appreciate this, consider an alternative derivation. Taking the curl of (1.11) we get

$$\frac{\partial\omega}{\partial t} = \nabla \times (\mathbf{v} \times \omega). \tag{1.20}$$

The simplest lesson one can immediately draw from (1.20) is that if $\omega \equiv 0$ then $\partial\omega/\partial t \equiv 0$ that is irrotational flow remains such in an ideal fluid. Not only zero but any value of vorticity is conserved for fluid particles. Indeed, (1.20) is the same equation that describes the magnetic field in a perfect conductor: Substituting the condition for the absence of the electric field in the frame moving with the velocity $\mathbf{v}$, $cE + \mathbf{v} \times H = 0$, into the Maxwell equation $\partial H/\partial t = -c\nabla \times E$, one gets $\partial H/\partial t = \nabla \times (\mathbf{v} \times H)$. The magnetic flux is conserved in a perfect conductor and so is the vorticity flux in an isentropic flow. One can visualize the vector field by introducing field lines, which give the direction of the field at any point while their density is proportional to the magnitude of the field. Kelvin's theorem means that vortex lines move with material elements in an inviscid fluid exactly like magnetic lines are frozen into a perfect conductor. One way to prove this is to show that ω/ρ (and H/ρ) satisfies the same equation as the distance $\mathbf{r}$ between two fluid particles: $\mathrm{d}\mathbf{r}/\mathrm{d}t = (\mathbf{r}\cdot\nabla)\mathbf{v}$. This is done using $\mathrm{d}\rho/\mathrm{d}t = -\rho\,\mathrm{div}\,\mathbf{v}$ and applying the general relation

$$\nabla \times (A \times B) = A(\nabla\cdot B) - B(\nabla\cdot A) + (B\cdot\nabla)A - (A\cdot\nabla)B \tag{1.21}$$

to $\nabla \times (\mathbf{v} \times \omega) = (\omega\cdot\nabla)\mathbf{v} - (\mathbf{v}\cdot\nabla)\omega - \omega\,\mathrm{div}\,\mathbf{v}$. We then obtain

$$\begin{aligned}
\frac{\mathrm{d}}{\mathrm{d}t}\frac{\omega}{\rho} &= \frac{1}{\rho}\frac{\mathrm{d}\omega}{\mathrm{d}t} - \frac{\omega}{\rho^2}\frac{\mathrm{d}\rho}{\mathrm{d}t} = \frac{1}{\rho}\left[\frac{\partial\omega}{\partial t} + (\mathbf{v}\cdot\nabla)\omega\right] + \frac{\omega\,\mathrm{div}\,\mathbf{v}}{\rho} \\
&= \frac{1}{\rho}[(\omega\cdot\nabla)\mathbf{v} - (\mathbf{v}\cdot\nabla)\omega - \omega\,\mathrm{div}\,\mathbf{v} + (\mathbf{v}\cdot\nabla)\omega] + \frac{\omega\,\mathrm{div}\,\mathbf{v}}{\rho} \\
&= \left(\frac{\omega}{\rho}\cdot\nabla\right)\mathbf{v}. \tag{1.22}
\end{aligned}$$

Since $\mathbf{r}$ and ω/ρ move together, then any two close fluid particles chosen on the vorticity line always stay on it. Consequently any fluid particle stays on the same vorticity line so that any fluid contour never crosses vorticity lines and the flux is indeed conserved. Compression transversal to the lines decreases the contour area thus increasing vorticity, similar effect leads to magnetohydrodynamic dynamo.

It is important to stress that Kelvin theorem is a nonlocal conservation law so it *is not* equivalent to the conservation of an angular momentum, which has a local density $\rho \mathbf{v} \times \mathbf{r}$. The symmetry which corresponds to the conservation of vorticity flux corresponds to re-labeling Lagrangian coordinates.

We have finished the formulations of the equations and their general properties and will now consider the simplest case that allows for an analytic study. This involves several assumptions.

1.2.4 Irrotational and incompressible flows

Irrotational flows are defined as having zero vorticity: $\omega = \nabla \times \mathbf{v} \equiv 0$. In such flows, $\oint \mathbf{v} \cdot d\mathbf{l} = 0$ round any closed contour, which means, in particular, that there are no closed streamlines for a singly connected domain. Note that the flow has to be isentropic to stay irrotational (i.e. inhomogeneous heating can generate vortices). A zero-curl vector field is potential, $\mathbf{v} = \nabla \phi$, so that the Euler equation (1.11) takes the form

$$\nabla \left(\frac{\partial \phi}{\partial t} + \frac{v^2}{2} + W \right) = 0.$$

After integration, one gets

$$\frac{\partial \phi}{\partial t} + \frac{v^2}{2} + W = C(t)$$

and the space-independent function $C(t)$ can be included into the potential, $\phi(r,t) \to \phi(r,t) + \int^t C(t') dt'$, without changing velocity. Eventually,

$$\frac{\partial \phi}{\partial t} + \frac{v^2}{2} + W = 0 . \tag{1.23}$$

For a steady flow, we thus obtained a more strong Bernoulli theorem with $v^2/2 + W$ being the same constant along all the streamlines, as distinct from a general case, where it may be a different constant along different streamlines.

Absence of vorticity provides for a dramatic simplification, which we exploit in this section and the next one. Unfortunately for pipeline operators and fortunately for birds, irrotational flows are much less frequent than Kelvin's theorem suggests. The main reason is that (even for isentropic flows) the viscous boundary layers near solid boundaries generate vorticity, as we shall see in Section 1.5. Yet we shall also see there that large regions of the flow can be unaffected by the vorticity generation and effectively described as irrotational. Another class of potential flows is provided by small-amplitude oscillations (like waves or motions due to oscillations of an immersed body). If the amplitude of oscillations a is small compared with the velocity scale of change l then $\partial v/\partial t \simeq v^2/a$ while $(v \nabla) v \simeq v^2/l$ so that the nonlinear term can be neglected and $\partial v/\partial t = -\nabla W$. Taking the curl of this equation we see that ω is conserved but its average is zero in oscillating motion so that $\omega = 0$.

After we simplified the Euler equation as much as possible, from (1.2) to (1.23), let us simplify the continuity equation.

Incompressible fluid can be considered as such if the density of any fluid element does not change in a flow: $d\ln\rho/dt = -\mathrm{div}\, v = 0$. For an incompressible fluid, the continuity equation is thus reduced to

$$\mathrm{div}\,\mathbf{v} = 0. \tag{1.24}$$

Strictly speaking, this could be true even when density varies in space and in time as long as any element keeps its density as it moves. In other words, $\partial\rho/\partial t$ and $(v\nabla)\rho$ can both be nonzero as long as their sum is zero. We, however, consider below only the simplest case when density is constant both in time and in space. This means that in the continuity equation, $\partial\rho/\partial t + (v\nabla)\rho + \rho\,\mathrm{div}\,v = 0$, the first two terms are much smaller than the third one. Let the velocity v change over the scale l and the time τ. The density variation can be estimated as

$$\delta\rho \simeq (\partial\rho/\partial p)_s \delta p \simeq (\partial\rho/\partial p)_s \rho v^2 \simeq \rho v^2/c^2, \tag{1.25}$$

where the pressure change was estimated from the Bernoulli relation. Requiring

$$(v\nabla)\rho \simeq v\delta\rho/l \ll \rho\,\mathrm{div}\,v \simeq \rho v/l,$$

we get the condition $\delta\rho \ll \rho$ which, according to (1.25), is true as long as the velocity is much less than the speed of sound:

For time-dependent flows, one must also require that the density changes slowly enough: $\partial\rho/\partial t \ll \rho\,\mathrm{div}\,v$. Comparing $\partial v/\partial t \simeq v/\tau$ and $\nabla p/\rho \simeq c^2\delta\rho/\rho l$, we estimate the density change due to temporal change as $\delta\rho \simeq l\rho v/\tau c^2$, so that

$$\partial\rho/\partial t \simeq \delta\rho/\tau \simeq l\rho v/\tau^2 c^2 \ll \rho\,\mathrm{div}\,v \simeq \rho v/l.$$

Therefore, the second condition of incompressibility is that the typical time of change τ must be much larger than the typical scale of change l divided by the sound velocity c:

$$\tau \gg l/c, \tag{1.26}$$

Indeed, sound equilibrates densities in different points so that all flow changes must be slow to let sound pass.

For isentropic motion of an incompressible fluid, the internal energy does not change ($dE = Tds + p\rho^{-2}d\rho$) so that one can put everywhere $W = p/\rho$.

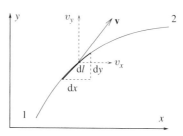

Figure 1.7 The flux through the line element dl is the flux to the right $v_x dy$ minus the flux up $v_y dx$, in agreement with (1.27).

Since density is no longer an independent variable, the equations that contain only velocity can be chosen: one takes (1.20) and (1.24).

In two dimensions, incompressible flow can be characterized by a single scalar function. Since $\partial v_x / \partial x = -\partial v_y / \partial y$ then we can introduce the *stream function* ψ defined by $v_x = \partial \psi / \partial y$ and $v_y = -\partial \psi / \partial x$. Recall that the streamlines are defined by $v_x dy - v_y dx = 0$, which now corresponds to $d\psi = 0$, that is the equation $\psi(x, y) = $ const. does indeed determine streamlines. Another important use of the stream function is that the flux through any line is equal to the difference of ψ at the endpoints (and is thus independent of the line form – an evident consequence of incompressibility):

$$\int_1^2 v_n dl = \int_1^2 (v_x dy - v_y dx) = \int d\psi = \psi_2 - \psi_1. \qquad (1.27)$$

Here v_n is the velocity projection on the normal; that is the flux is equal to the modulus of the vector product $\int |\mathbf{v} \times \mathbf{dl}|$, see Figure 1.7. A solid boundary at rest has to coincide with one of the streamlines.

Potential flow of an incompressible fluid. By virtue of (1.24) the potential satisfies the Laplace equation[6]

$$\Delta \phi = 0 \,,$$

with the condition $\partial \phi / \partial n = 0$ on a solid boundary at rest.

It is the distinctive property of an irrotational incompressible flow that the velocity distribution is defined completely by a *linear* equation. Owing to linearity, velocity potentials can be superimposed (but not pressure distributions). Before considering particular flows, we formulate several general statements. First, the Laplace equation is elliptic, which means that the solutions are smooth inside the domains, singularities could exist on boundaries only, in contrast to hyperbolic (say, wave) equations. Remind that the second-order linear differential operator $\sum a_i \partial_i^2$ is called elliptic if all a_i are of the same sign, hyperbolic if their signs are different and parabolic if at least one coefficient is zero. The names come from the fact that a real quadratic curve $ax^2 + 2bxy + cy^2 = 0$ is a hyperbola, an ellipse or a parabola depending on whether $ac - b^2$ is negative, positive or zero. For hyperbolic equations, one can introduce characteristics where solution stays constant; if different characteristics cross then a singularity may appear inside the domain. Solutions of elliptic equations are smooth, their stationary points are saddles rather than maxima or minima. See also Sections 2.3.2 and 2.3.5.

Second, integrating (1.29) over any volume one gets

$$\int \Delta\phi\, dV = \int \text{div}\nabla\phi\, dV = \oint \nabla\phi \cdot d\mathbf{f} = 0,$$

that is the flux is zero through any closed surface (as is expected for an incompressible fluid). That means, in particular, that $\mathbf{v} = \nabla\phi$ changes sign on any closed surface so that extrema of ϕ could be on the boundary only. The same can be shown for velocity components (e.g. for $\partial\phi/\partial x$) since they also satisfy the Laplace equation. That means that for any point P inside one can find P' having higher $|v_x|$. If we choose the x-direction to coincide with $\nabla\phi$ at P we conclude that for any point inside one can find another point in the immediate neighborhood where $|v|$ is greater. In other terms, v^2 cannot have a maximum inside (but can have a minimum). Similarly for pressure, taking the Laplacian of the Bernoulli relation (1.30),

$$\Delta p = -\rho\Delta v^2/2 = -\rho(\nabla v)^2,$$

and integrating it over volume, one obtains

$$\oint \nabla p \cdot d\mathbf{f} = -\rho \int (\nabla v)^2 dV < 0,$$

that is a pressure minimum could be only on a boundary (although a maximum can occur at an interior point). For steady flows, $v^2/2 + p/\rho = $ const. so that the points of max v^2 coincide with those of min p and all are on a boundary.[7] The knowledge of points of minimal pressure is important for cavitation, which is a creation of gas bubbles when the pressure falls below the vapor pressure; when such bubbles then experience higher pressure, they may collapse, producing shock waves that cause severe damage to moving boundaries like turbine blades and ships' propellers. Shock are also created when the local fluid velocity exceeds the velocity of sound, as we shall see in Section 2.3.2; this must happen first at the velocity maxima, which again are possible only on the boundary of a potential flow.

Despite the high degree of idealization, the theory of incompressible potential flows is of significant practical importance. Not only it describes large regions of the flows outside wakes past the bodies, as described in Section 1.5.4, but also high-power explosions as described in Exercise 1.18.

Two-dimensional case. Particularly beautiful is the description of two-dimensional (2D) potential incompressible flows. Both potential and stream function exist in this case. The equations

$$v_x = \frac{\partial \phi}{\partial x} = \frac{\partial \psi}{\partial y}, \qquad v_y = \frac{\partial \phi}{\partial y} = -\frac{\partial \psi}{\partial x}, \tag{1.28}$$

could be recognized as the Cauchy–Riemann conditions for the complex potential $w = \phi + i\psi$ to be an analytic function of the complex argument $z = x + iy$. That means that the rate of change of w does not depend on the direction in the x, y-plane, so that one can define the complex derivative dw/dz, which exists everywhere. For example, both choices $dz = dx$ and $dz = idy$ give the same answer by virtue of (1.28):

$$\frac{dw}{dz} = \frac{\partial \phi}{\partial x} + i\frac{\partial \psi}{\partial x} = \frac{\partial \phi}{i\partial y} + \frac{\partial \psi}{\partial y} = v_x - iv_y = ve^{-i\theta}, \quad \mathbf{v} = v_x + iv_y = \frac{d\bar{w}}{d\bar{z}}.$$

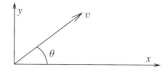

The complex form allows one to describe many flows in a compact form and find flows in a complex geometry by mapping a domain onto a standard one. Such a transformation must be conformal, i.e. done using an analytic function so that the equations (1.28) preserve their form in the new coordinates. Conformal transformations stretch uniformly in all directions at every point but the magnitude of stretching generally depends on a point. As a result, conformal maps preserve angles but not the distances. These properties had been first made useful in naval cartography (Mercator 1569) well before the invention of the complex analysis. Indeed, to discover a new continent it is preferable to know the direction rather than the distance ahead.

We thus get our first (infinite) family of flows: any complex function analytic in a domain and having a constant imaginary part on the boundary describes a potential flow of an incompressible fluid in this domain. Uniform flow is just $w = (v_x - iv_y)z$. Here are two other examples:

(1) Potential flow near a stagnation point $\mathbf{v}=0$ (inside the domain or on a
    smooth boundary) is expressed via the rate-of-strain tensor S_{ij}:
    $\phi = S_{ij}x_ix_j/2$ with $\operatorname{div}\mathbf{v}=S_{ii}=0$. In the principal axes of the tensor, one
    has $v_x=kx$, $v_y=-ky$, which corresponds to

$$\phi = k(x^2-y^2)/2, \quad \psi=kxy, \quad w=kz^2/2.$$

The streamlines $x=\psi/ky$ and trajectories $x(t)=x(0)y(0)/y(t)$ are
rectangular hyperbolae. This is applied, in particular, on the boundary,
which has to coincide with one of the principal axes (x or y) or both. The
figure presents the flows near the boundary along x and along x and y
(half of the previous one):

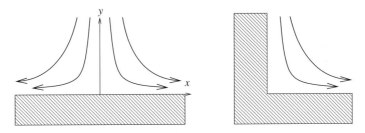

(2) Consider the potential in the form $w=Az^n$, that is $\phi=Ar^n\cos n\theta$ and
    $\psi=Ar^n\sin n\theta$. Zero-flux boundaries should coincide with the streamlines
    so two straight lines, $\theta=0$ and $\theta=\pi/n$ could be seen as boundaries.
    Choosing different n, one can have different interesting particular cases.
    The velocity modulus

$$v=\left|\frac{dw}{dz}\right|=n|A|r^{n-1}$$

at $r\to0$ either turns to 0 ($n>1$) or to ∞ ($n<1$) (Figure 1.8).

One can think of those solutions as obtained by a conformal transformation $\zeta=z^n$, which maps the z-domain into the full ζ-plane. The potential

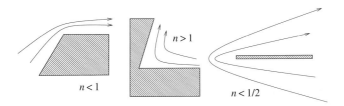

Figure 1.8 Flows described by the complex potential $w=Az^n$.

$w = Az^n = A\zeta$ describes a uniform flow in the ζ-plane. Respective z and ζ points have the same value of the complex potential so that the transformation maps streamlines onto streamlines. The velocity in the transformed domain is $\mathrm{d}w/\mathrm{d}\zeta = (\mathrm{d}w/\mathrm{d}z)(\mathrm{d}z/\mathrm{d}\zeta)$, that is the velocity modulus is inversely proportional to the stretching factor of the transformation. This has two important consequences: first, the energy of the potential flow is invariant with respect to conformal transformations, i.e. the energy inside every closed curve in the z-plane is the same as the energy inside the image of the curve in the ζ-plane. Second, the flow dynamics are not conformal invariant even along conformal invariant streamlines (which coincide with particle trajectories for a steady flow). Indeed, when the flow shifts the fluid particle from z to $z + v\mathrm{d}t = z + \mathrm{d}t(\mathrm{d}\bar{w}/\mathrm{d}\bar{z})$, the new image,

$$\zeta(z + v\mathrm{d}t) = \zeta(z) + \mathrm{d}t\, v\frac{\mathrm{d}\zeta}{\mathrm{d}z} = \zeta(z) + \mathrm{d}t\frac{\mathrm{d}\bar{w}}{\mathrm{d}\bar{z}}\frac{\mathrm{d}\zeta}{\mathrm{d}z},$$

does not coincide with the new position of the old image,

$$\zeta(z) + \mathrm{d}t\frac{\mathrm{d}\bar{w}}{\mathrm{d}\bar{\zeta}} = \zeta(z) + \mathrm{d}t\frac{\mathrm{d}\bar{w}}{\mathrm{d}\bar{z}}\frac{\mathrm{d}\bar{z}}{\mathrm{d}\bar{\zeta}}\ .$$

Despite the beauty of conformal flows, their applications are limited. Real flow usually separates at discontinuities; it does not turn over the corner for $n < 1$ and does not reach the inside of the corner for $n > 1$:

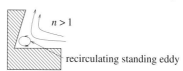

$n > 1$

recirculating standing eddy

The phenomenon of separation is due to a combined action of friction and inertia and is discussed in detail in Section 1.5.2. Separation produces vorticity, which makes it impossible to introduce the potential ϕ and use the complex potential w (streamlines of rotational flows are not conformal invariant).

1.3 Moving through fluids

How much force one needs to set a body in motion through a fluid? Consider an air bubble in champagne. The mass of the air in the bubble is thousand times less than the mass of the displaced liquid which determines the Archimedes force. Will the bubbles then speed up with the acceleration of $1000g$, literally blowing the liquid into our face? Indeed, a moving body must involve a certain amount of moving fluid. We then are tempted to conclude that the force acting on the body must be equal to the time derivative of the momentum of the body *plus* the momentum of the fluid. Reflecting a bit more, we see that this is incorrect too: if, for instance, the liquid is enclosed by rigid walls then its momentum is identically zero! But if those walls are far away from the body, they must not influence the force. We then expect that the body sets in motion some fluid in the vicinity (and this determines the force), while the compensating reflux is spread over the whole fluid and does not influence the force. In this section, we describe the flows of ideal incompressible fluid set in motion by moving bodies. We start from the most symmetric case of a moving sphere and then consider a body of arbitrary shape. From each flow, we shall derive the contribution of the fluid into the force needed to set the body in motion. That contribution is equivalent to an appearance of added mass, apparently the first example of renormalization in physics. We shall find out that it is the quasimomentum (not momentum) of the fluid that determined the added mass. We discuss the difference between the conservation of momentum (due to space homogeneity) and quasimomentum (due to fluid homogeneity).

For nonsymmetric moving bodies, fluid generally applies force not only opposite to the direction of motion but also across. The force perpendicular to the motion is called lift, since it keeps birds and planes from falling from the skies. In this section, we find out that there is neither lift nor drag for a steady motion in an ideal fluid which will induce us to introduce friction in the next section.

1.3.1 Incompressible potential flow past a body

If we assume that the flow appears from rest, Kelvin theorem suggests that the vorticity must be identically zero. Alternatively, in the reference frame of the body, fluid comes from infinity, where a uniform flow also has zero vorticity. Flow is thus assumed to be four "i": infinite, irrotational, incompressible and ideal. The algorithm to describe such a flow is to solve the Laplace equation

$$\Delta \phi = 0. \tag{1.29}$$

Solutions of the equation $\Delta \phi = 0$ that vanish at infinity are $1/r$ and its derivatives, $\partial^n (1/r)/\partial x^n$, in three dimensions. In two dimensions the solutions are $log(r)$ and its derivatives.

The boundary condition on the body surface is the requirement that the normal components of the body and fluid velocities coincide, that is at any given moment one has $\partial \phi/\partial n = u_n$, where $\mathbf{u}$ is the body velocity. After finding the potential, one calculates $\mathbf{v} = \nabla \phi$ and then finds pressure from the Bernoulli equation:

$$p = -\rho(\partial \phi/\partial t + v^2/2). \tag{1.30}$$

1.3.2 Moving sphere

Owing to the complete symmetry of the sphere, its motion is characterized by a single vector of its velocity $\mathbf{u}$. Linearity requires $\phi \propto \mathbf{u}$ so the flow potential could be only made as a scalar product of the vectors $\mathbf{u}$ and the gradient, which is the dipole field:

$$\phi = a(\mathbf{u} \cdot \nabla)\frac{1}{r} = -a\frac{(\mathbf{u} \cdot \mathbf{n})}{r^2}$$

where $\mathbf{n} = \mathbf{r}/r$. On the body, $r = R$ and $\mathbf{v} \cdot \mathbf{n} = \mathbf{u} \cdot \mathbf{n} = u\cos\theta$. Using $\phi = -ua\cos\theta/r^2$ and $v_R = 2auR^{-3}\cos\theta$, this condition gives $a = R^3/2$.

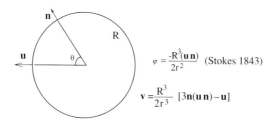

$$\varphi = \frac{-R^3(\mathbf{u}\,\mathbf{n})}{2r^2} \quad \text{(Stokes 1843)}$$

$$\mathbf{v} = \frac{R^3}{2r^3}\,[3\mathbf{n}(\mathbf{u}\,\mathbf{n}) - \mathbf{u}]$$

Now one can calculate the pressure

$$p = p_0 - \rho v^2/2 - \rho\partial\phi/\partial t ,$$

having in mind that our solution moves with the sphere, that is $\phi(\mathbf{r} - \mathbf{u}t, \mathbf{u})$ and

$$\frac{\partial\phi}{\partial t} = \dot{\mathbf{u}} \cdot \frac{\partial\phi}{\partial \mathbf{u}} - \mathbf{u} \cdot \nabla\phi ,$$

which gives

$$p = p_0 + \rho u^2 \frac{9\cos^2\theta - 5}{8} + \frac{\rho R}{2}\mathbf{n} \cdot \dot{\mathbf{u}} .$$

The force acting on the body is minus the pressure integral over the surface $-\oint p\,d\mathbf{f}$. For example, assuming that the body velocity does not change its direction, $\mathbf{n} \cdot \dot{\mathbf{u}} = \dot{u}\cos\theta$, we derive

$$F_x = -\oint p\cos\theta\,df = -\rho R^3\dot{u}\pi \int \cos^2\theta\,d\cos\theta = -2\pi\rho R^3\dot{u}/3 . \quad (1.31)$$

We thus see that the mass of the fluid one needs to accelerate is half the mass of the displaced fluid. This is one of the simplest examples of renormalization in physics: the body moving through a fluid acquires additional (also called induced) mass. A spherical air bubble in a liquid has a mass that is half of the mass of the displaced liquid; since the buoyancy force is the displaced mass times g then we conclude with a relief that the initial bubble acceleration is close to $2g$. As the bubble accelerates relative to the liquid, the drag force increases, as will be described in Section 1.5, and the acceleration decreases.

If the radius depends on time too then $F_x \propto \partial\phi/\partial t \propto -\partial(R^3 u)/\partial t$. Remarkably, it means that a shrinking moving body experiences an accelerating force, this will be discussed after (1.44).

According to our formulas, for a uniformly moving sphere with a constant radius, $\dot{R} = \dot{\mathbf{u}} = 0$, the force is zero: $\oint p\,d\mathbf{f} = 0$. In a two-dimensional case (say, flow around a cylinder with the radius R), the potential is $\phi = -R^2(\mathbf{u} \cdot \nabla)\log r$ and the pressure on the surface is $p = -2\rho \sin^2\theta + \rho R\mathbf{n} \cdot \dot{\mathbf{u}}$. Similar to (1.31), after angular integration only the last term contributes the total force which thus vanishes for a steady motion. The force is zero because of fore-and-aft symmetry: incoming and outgoing parts of the steady flow are identical and provide for the same pressure fields. This flies in the face of our common experience: fluids do resist bodies even moving with at a constant speed. Maybe we obtained zero force in a steady case due to a symmetrical shape of the body?

1.3.3 Moving body of an arbitrary shape

In two dimensions, potential flow around a body with an arbitrary shape can be obtained by a conformal map of the solution for a circle. If a steady motion of a circle meets no force, the same is true for any body shape. Indeed, for a steady flow, the pressure (up to a constant) is $p = -\rho v^2/2 = -\rho|dw/dz|^2/2$. One can combine the forces into the integral over the body surface where $dw = d\bar{w}$ (Blasius 1910):

$$F_x - iF_y = -i\oint p\,d\bar{z} = \frac{i\rho}{2}\oint \frac{dw}{dz}\,d\bar{w} = \frac{i\rho}{2}\oint \left(\frac{dw}{dz}\right)^2 dz.$$

We integrate an analytic function having no singularities and can enlarge the contour to any extent, so that the integral vanishes, since velocity decreases faster than $1/|z|$ due to mass conservation. Note that this consideration does not rule out nonzero torque acting on a body.

In three dimensions, flow past a body of an arbitrary shape generally cannot be found analytically. However, the main beauty of the potential theory (and conformal analysis used in the previous paragraph) is that one can say something about "here" by considering the field "there." In our case, we are interested in the forces acting on the body surface yet we consider the flow far away, which must be weakly dependent on the body shape. Indeed, at large distances from the body, a solution of $\Delta\phi = 0$ is again sought in the form of the first non-vanishing multipole. The first (charge) term $\phi = a/r$ cannot be present because it corresponds to the velocity $\mathbf{v} = -a\mathbf{r}/r^3$ with the radial component $v_r = a/r^2$ providing for an r-independent flux $4\pi\rho a$ through a closed sphere of radius r; existence of a flux contradicts mass conservation. So the first nonvanishing term is again a dipole:

$$\phi = \mathbf{A} \cdot \nabla(1/r) = -(\mathbf{A} \cdot \mathbf{n})r^{-2},$$
$$\mathbf{v} = [3(\mathbf{A} \cdot \mathbf{n})\mathbf{n} - \mathbf{A}]r^{-3}.$$

For the sphere above, $\mathbf{A} = \mathbf{u}R_0^3/2$, where R_0 is the radius. For non-symmetric bodies, the vectors $\mathbf{A}$ and $\mathbf{u}$ are not collinear, though linearly related $A_i = \alpha_{ik}u_k$, where the tensor α_{ik} (having the dimensionality of volume) depends on the body shape.

To relate the force acting on the body to the flow at large distances, let us start by calculating the energy $E = \rho \int v^2 \, dV/2$ of the moving fluid outside the body and inside the large sphere of radius R. We present $v^2 = u^2 + (\mathbf{v} - \mathbf{u})(\mathbf{v} + \mathbf{u})$ and write $\mathbf{v} + \mathbf{u} = \nabla(\phi + \mathbf{u} \cdot \mathbf{r})$. Using $\operatorname{div} v = \operatorname{div} u = 0$ one can write

$$\int_{r<R} v^2 \, dV = u^2(V - V_0) + \int_{r<R} \operatorname{div}[(\phi + \mathbf{u} \cdot \mathbf{r})(\mathbf{v} - \mathbf{u})] \, dV$$
$$= u^2(V - V_0) + \oint_{S+S_0} (\phi + \mathbf{u} \cdot \mathbf{r})(\mathbf{v} - \mathbf{u}) \, d\mathbf{f}$$
$$= u^2(V - V_0) + \oint_{S} (\phi + \mathbf{u} \cdot \mathbf{r})(\mathbf{v} - \mathbf{u}) \, d\mathbf{f}.$$

Substituting

$$\phi = -(\mathbf{A} \cdot \mathbf{n})R^{-2}, \qquad \mathbf{v} = [3\mathbf{n}(\mathbf{A} \cdot \mathbf{n}) - \mathbf{A}]R^{-3}$$

and integrating over angles,

$$\int (\mathbf{A} \cdot \mathbf{n})(\mathbf{u} \cdot \mathbf{n}) \, d\Omega = A_i u_k \int n_i n_k \, d\Omega = A_i u_k \delta_{ik} \int \cos^2\theta \sin\theta \, d\theta d\varphi$$
$$= (4\pi/3)(\mathbf{A} \cdot \mathbf{u}),$$

we obtain the energy in the form

$$E = \rho[4\pi(\mathbf{A} \cdot \mathbf{u}) - V_0 u^2]/2 = m_{ik}u_i u_k/2. \tag{1.32}$$

Here we introduce the *induced-mass tensor*:

$$m_{ik} = 4\pi\rho\,\alpha_{ik} - \rho V_0 \delta_{ik}.$$

For a sphere, $m_{ik} = \rho V_0 \delta_{ik}/2$, that is half the mass of the displaced fluid. Induced mass can be much larger (for a thin disc moving perpendicular to its plane) and much smaller (for a needle moving "end on") than the displaced mass.

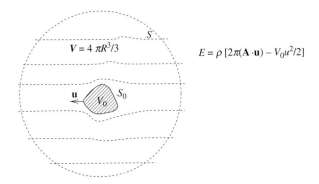

We now have to pass from the energy to the force acting on the body which is done by considering the change in the energy of the body (the same as minus the change of the fluid energy dE) being equal to the work done by force F on the path udt: $dE = -\mathbf{F} \cdot \mathbf{u}dt$. The change of the momentum of the body is $d\mathbf{P} = -\mathbf{F}dt$ so that $dE = \mathbf{u} \cdot d\mathbf{P}$. This relation is true for changes caused by the velocity change by force (not by the change in the body shape) so that the change of the body momentum is $dP_i = m_{ik}du_k$ and the force is

$$F_i = -m_{ik}\dot{u}_k. \tag{1.33}$$

We see that the presence of potential flow means only an additional mass but not resistance for an arbitrary body shape. The reason is that a steady potential flow in an incompressible fluid has fore-and-aft symmetry at infinity so that the total momentum does not change inside the volume containing the body.

How to generalize (1.33) for the case when both m_{ik} and $\mathbf{u}$ change? Our consideration for a sphere suggests that, since the respective contribution into the pressure is $-\rho \partial\phi/\partial t \propto \partial(m_{ik}u_in_k)/\partial t$, then the proper generalization is (density is assumed time-independent)

$$F_i = -\frac{d}{dt}m_{ik}u_k. \tag{1.34}$$

It looks as if $m_{ik}u_k$ is the momentum of the fluid yet it is not (it is quasi-momentum as explained in the next section).[8]

The equation of motion for the body under the action of an external force $\mathbf{f}$,

$$\frac{d}{dt}Mu_i = f_i + F_i = f_i - \frac{d}{dt}m_{ik}u_k,$$

could be written in a form that makes the term induced mass clear:

$$\frac{d}{dt}(M\delta_{ik} + m_{ik})u_k = f_i. \tag{1.35}$$

Remark on the case of time-dependent body mass $M(t)$ is in order. Was it correct to put the body mass inside the time derivative? The answer depends on whether the body mass grows due to condensation (as in the problem 1.14) or decreases due to evaporation or dissolution (as in the problem 2.5). In the former case, if the vapour was at rest before condensation, then it adds no momentum after condensation. Therefore, the momentum change is only due to the external force, so that the time derivative of the momentum is equal to the force acting on the body, as in (1.35). In the latter case, the material leaves with the same velocity, that is evaporation by itself does not change the body velocity. The velocity change is due to the force, so that the velocity time derivative must be equal to the force divided by the mass, and the equation takes the form:

$$M\frac{d}{dt}u_i + \frac{d}{dt}m_{ik}u_k = f_i. \qquad (1.36)$$

Body in a flow. Consider now an opposite situation when the fluid moves in an oscillating way while a small body is immersed in the fluid. For example, a long sound wave propagates in a fluid. We do not consider here the external forces that move the fluid; we wish to relate the body velocity $\mathbf{u}$ to the fluid velocity $\mathbf{v}$, which is supposed to be homogeneous on the scale of the body size. If the body moved with the same velocity, $\mathbf{u} = \mathbf{v}$, then it would be under the action of a force that would act on the fluid in its place, $\rho V_0 \dot{\mathbf{v}}$. Relative motion gives the reaction force $dm_{ik}(v_k - u_k)/dt$. The sum of the forces gives the time derivative of the body momentum:

$$\frac{d}{dt}Mu_i = \rho V_0 \dot{v}_i + \frac{d}{dt}m_{ik}(v_k - u_k). \qquad (1.37)$$

Integrating over time with the integration constant zero (since $u=0$ when $v=0$) we get the relation between the velocities of the body and of the fluid:

$$(M\delta_{ik} + m_{ik})u_k = (m_{ik} + \rho V_0 \delta_{ik})v_k.$$

For a sphere, $\mathbf{u} = \mathbf{v}3\rho/(\rho + 2\rho_0)$, where ρ_0 is the density of the body. For a liquid droplet in air, $u = 3v\rho/2\rho_0 \ll v$. For an air bubble in a liquid, $\rho_0 \ll \rho$ and $u \approx 3v$. The motion of denser/lighter bodies is retarded/advanced relative to the fluid.

1.3.4 Quasimomentum and induced mass

In the previous section, we obtained the force acting on an accelerating body via the energy of the fluid and the momentum of the body because the momentum of the fluid, $\mathbf{M} = \rho \int \mathbf{v} \, dV$, is not well-defined for a potential flow around the body. For example, the integral of $v_x = D(3\cos^2\theta - 1)r^{-3}$ depends on the form of the volume chosen: it is zero for a spherical volume and nonzero for a cylinder of length L and radius $\mathscr{R}$ set around the body:

$$\int_{-1}^{1} (3\cos^2\theta - 1)\,d\cos\theta = 0,$$

$$M_x = 4\pi\rho D \int_{-L}^{L} dz \int_{0}^{\mathscr{R}} r\,dr \frac{2z^2 - r^2}{(z^2 + r^2)^{5/2}} = \frac{4\pi\rho DL}{(L^2 + \mathscr{R}^2)^{1/2}} \, . \tag{1.38}$$

Does that mean that the momentum stored in the fluid depends on the boundary conditions at infinity? In fact, it does. For example, the motion by a sphere in a fluid enclosed by rigid walls must be accompanied by the displacement of an equal amount of fluid in the opposite direction; then the momentum of the fluid must be $-\rho V_0 u$ rather than $\rho u V_0/2$. The negative momentum $-3\rho u V_0/2$ delivered by the walls is absorbed by the whole body of fluid and results in an infinitesimal back-flow, while the momentum $\rho u V_0/2$ delivered by the sphere results in a finite localized flow. From (1.38) we can get a shape-independent answer $4\pi\rho D$ only in the limit $L/\mathscr{R} \to \infty$. To recover the answer $4\pi\rho D/3$ ($= \rho u V_0/2 = \rho u 2\pi R^3/3$ for a sphere) that we expect from (1.34), we need to subtract the reflux $8\pi\rho D/3 = \rho u 4\pi R^3/3$, compensating the body motion.[9]

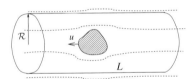

It is the quasimomentum of the fluid particles that is independent of the remote boundary conditions and whose time derivative gives the inertial force (1.34) acting on the body. Conservation laws of the momentum and the quasimomentum follow from different symmetries. The momentum expresses invariance of the Hamiltonian $\mathscr{H}$ with respect to the shift of coordinate system. If the space is filled by a medium (fluid or solid), then the quasimomentum expresses invariance of the Hamiltonian with respect to a space shift, *keeping the medium fixed*. That invariance follows from the identity of different elements of the medium. In a crystal, such shifts are allowed only by the lattice spacing. In a continuous medium, shifts are arbitrary. In this case, the system Hamiltonian must be independent of the coordinates:

$$\frac{\partial\mathscr{H}}{\partial x_i} = \frac{\partial\mathscr{H}}{\partial\pi_j}\frac{\partial\pi_j}{\partial x_i} + \frac{\partial\mathscr{H}}{\partial q_j}\frac{\partial q_j}{\partial x_i} = 0, \tag{1.39}$$

where the vectors $\pi(\mathbf{x},t), q(\mathbf{x},t)$ are respectively canonical momentum and coordinates (in every point in space). We need to define the quasimomentum $\mathbf{K}$ whose conservation is due to invariance of the Hamiltonian: $\partial K_i/\partial t = \partial \mathcal{H}/\partial x_i = 0$. Recall that the time derivative of any function of canonical variables is given by the Poisson bracket of this function with the Hamiltonian:

$$\frac{\partial K_i}{\partial t} = \{K_i, \mathcal{H}\} = \frac{\partial K_i}{\partial q_j}\frac{\partial \mathcal{H}}{\partial \pi_j} - \frac{\partial K_i}{\partial \pi_j}\frac{\partial \mathcal{H}}{\partial q_j}$$
$$= \frac{\partial \mathcal{H}}{\partial x_i} = \frac{\partial \mathcal{H}}{\partial \pi_j}\frac{\partial \pi_j}{\partial x_i} + \frac{\partial \mathcal{H}}{\partial q_j}\frac{\partial q_j}{\partial x_i},$$

This gives the partial differential equations for the quasimomentum,

$$\frac{\partial K_i}{\partial \pi_j} = -\frac{\partial q_j}{\partial x_i}, \quad \frac{\partial K_i}{\partial q_j} = \frac{\partial \pi_j}{\partial x_i},$$

whose solution is as follows:

$$K_i = -\int d\mathbf{x}\pi_j \frac{\partial q_j}{\partial x_i}. \tag{1.40}$$

For isentropic (generally compressible) flow of an ideal fluid, the Hamiltonian description can be given in Lagrangian coordinates, which describe the current position of a fluid element (particle) $\mathbf{r}(\mathbf{R},t)$ as a function of its initial position $\mathbf{R}$ and time t. Since we want our variable to have a finite change for a localized flow, we choose the canonical coordinate as the displacement $\mathbf{q} = \mathbf{r} - \mathbf{R}$, which is the continuum limit of the variable that describes lattice vibrations in the solid state physics. The canonical momentum is $\pi(\mathbf{R},t) = \rho_0(\mathbf{R})\mathbf{v}(\mathbf{R},t)$ where the velocity is $\mathbf{v} = (\partial \mathbf{r}/\partial t)_\mathbf{R} \equiv \dot{\mathbf{r}}$. Here ρ_0 is the density in the reference (initial) state, which can always be chosen to be uniform. As we discussed in Section 1.1.2 deriving the Euler equation, the fluid particle always has the same (unit) mass.

The Hamiltonian is as follows:

$$\mathcal{H} = \int \rho_0 \left[W(\mathbf{q}) + v^2/2\right] d\mathbf{R}, \tag{1.41}$$

where $W = E + p/\rho$ is the enthalpy, which thus plays a role of the potential energy of a fluid particle. The Hamiltonian for a fluid particle, $W + v^2/2$, is generally time-dependent.[10] The density is as follows: $\rho(\mathbf{R},t) = \rho_0\det[\partial r_i/\partial R_j]$.

Canonical equations of motion, $\dot{q}_i = \partial \mathcal{H}/\partial \pi_i$ and $\dot{\pi}_i = -\partial \mathcal{H}/\partial q_i$, give, respectively, $\dot{r}_i = v_i$ and $\dot{v}_i = -\partial W/\partial r_i = -\rho^{-1}\partial p/\partial r_i$. The velocity $\mathbf{v}$ is now an independent variable and not a function of the coordinates $\mathbf{r}$. All the time derivatives are for fixed $\mathbf{R}$, i.e. they are substantial derivatives. The quasimomentum (1.40) is as follows:

$$K_i = -\rho_0 \int v_j \frac{\partial q_j}{\partial R_i}\, d\mathbf{R} = \rho_0 \int v_j \left(\delta_{ij} - \frac{\partial r_j}{\partial R_i}\right) d\mathbf{R}\,, \qquad (1.42)$$

In plain words, only those particles contribute quasimomentum whose motion is disturbed by the body, so that for them $\partial r_j/\partial R_i \neq \delta_{ij}$. The integral (1.42) converges for spatially localized flows since $\partial r_j/\partial R_i \to \delta_{ij}$ when $R \to \infty$. Unlike (1.38), the quasimomentum (1.42) is independent of the form of distant boundaries. Using $\rho_0 d\mathbf{R} = \rho\, d\mathbf{r}$ one can also present

$$\begin{aligned}
K_i &= \rho_0 \int v_j \left(\delta_{ij} - \frac{\partial r_j}{\partial R_i}\right) d\mathbf{R} \\
&= \int \rho v_i\, d\mathbf{r} - \rho_0 \int v_j \frac{\partial r_j}{\partial R_i}\, d\mathbf{R}\,,
\end{aligned} \qquad (1.43)$$

i.e. indeed the quasimomentum is the momentum minus what can be interpreted as a reflux.

The conservation can now be established, substituting the equation of motion $\rho \dot{\mathbf{v}} = -\partial p/\partial \mathbf{r}$ into

$$\begin{aligned}
\dot{K}_i &= -\rho_0 \int \left(\dot{v}_j \frac{\partial q_j}{\partial R_i} + v_j \frac{\partial v_j}{\partial R_i}\right) d\mathbf{R} \\
&= -\rho_0 \int \left[\dot{v}_j \left(\frac{\partial r_j}{\partial R_i} - \delta_{ij}\right) + v_j \frac{\partial v_j}{\partial R_i}\right] d\mathbf{R} \\
&= -\rho_0 \int \left(\frac{\delta_{ij}}{\rho} \frac{\partial p}{\partial r_j} - \frac{\partial W}{\partial r_j} \frac{\partial r_j}{\partial R_i} + \frac{\partial v^2}{2 \partial R_i}\right) d\mathbf{R} \\
&= -\int \frac{\partial p}{\partial r_i}\, d\mathbf{r} + \int \frac{\partial}{\partial R_i}\left(W - \frac{\rho_0 v^2}{2}\right) d\mathbf{R} \\
&= -\int \frac{\partial p}{\partial r_i}\, d\mathbf{r} = \oint p\, df_i\,.
\end{aligned} \qquad (1.44)$$

In the fourth line, the integral over the reference space $\mathbf{R}$ of the total derivative in the second term is identical to zero, while the integral over $\mathbf{r}$ in the first term excludes the volume of the body, so that the boundary term remains, which is minus the force acting on the body. Therefore, the sum of the quasimomentum of the fluid and the momentum of the body is conserved in an ideal fluid. That explains the surprising effect of acceleration of a shrinking moving body. Indeed, when the induced mass and the quasimomentum of the fluid decrease then the body momentum must increase. Swimmers can get extra acceleration proportional to minus the time derivative of the added mass by a rapid decrease of the cross-section. Breaststroke kick-glide transition better be fast. The octopus is particularly effective in rapid shape-change during acceleration.

This quasimomentum (1.42) is defined for any flow. For a potential incompressible flow, one can obtain quasimomentum simply integrating the potential over the body surface: $\mathbf{K} = -\int \rho\phi\, d\mathbf{f}$. Indeed, consider a very short and strong pulse of pressure needed to bring the body from rest into motion, formally $p \propto \delta(t)$. During the pulse, the body doesn't move, so its position and surface are well-defined. In the Bernoulli relation (1.23) one can then neglect the v^2 term:

$$\frac{\partial\phi}{\partial t} = -\frac{v^2}{2} - \frac{p}{\rho} \approx -\frac{p}{\rho}. \tag{1.45}$$

Integrating the relation $-\rho\phi = \int p(t)\, dt$ over the body surface we get minus the change of the body momentum, i.e. the quasimomentum of the fluid. For example, integrating $\phi = -R^3 u\cos\theta/2r^2$ over the sphere we get

$$K_x = -\int \rho\phi\cos\theta\, d\mathbf{f} = 2\pi\rho R^3 u \int_{-1}^{1} \cos^2\theta\, d\cos\theta = 2\pi\rho R^3 u/3,$$

as expected. The difference between momentum and quasimomentum can be related to the momentum flux across the infinite surface due to pressure, which decreases as r^{-2} for a potential flow.

The quasimomentum of the fluid is related to the body velocity via the induced mass, $K_i = m_{ik}u_k$, so that one can use (1.42) to evaluate m_{ik}. For this, one needs to solve the Lagrangian equation of motion $\dot{\mathbf{r}} = \mathbf{v}(\mathbf{r},t)$; then one can show that the induced mass can be associated with the displacement of the fluid after the body pass. That would be wrong to think that in moving toward the right the body would displace fluid in front and leave vacant space behind, so that there should be a net reflux toward the left in compensation. Let us show that net displacement is always in the direction of the motion. Consider an arbitrary potential flow shifting with a constant speed, $\phi(x - ut, y, z)$, either due to moving body or propagating wave. To express the fluid displacement via the Lagrangian integral, we write

$$v_x = \frac{\partial\phi}{\partial x} = -\frac{1}{u}\frac{\partial\phi}{\partial t} = \frac{1}{u}\left(|\nabla\phi|^2 - \frac{d\phi}{dt}\right)$$

and substitute

$$x(t) - x(0) = \int_0^t v_x\big(x(t'), y, z\big)\, dt' = \frac{1}{u}\int_0^t |\nabla\phi|^2\, dt - \frac{1}{u}\phi|_0^t. \tag{1.46}$$

We now choose t such that the last term is zero (period for the wave or infinity for passing body) and see that the displacement is always positive.

The body pushes the fluid in front and pulls the fluid behind, while the fluid on the sides moves opposite to the body. Indeed, the potential is

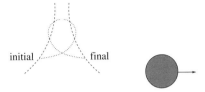

Figure 1.9 Displacement of the fluid by the passing body. The dotted line is the trajectory of the fluid particle. Two broken lines (chosen symmetrical) show the initial and final positions of the particles before and after the passage of the body.

$\phi \propto -x/r^d$ (where $d = 3$ for a sphere, $d = 2$ for a cylinder), so that the horizontal component of the fluid velocity, $v_x \propto d\cos^2\theta - 1$, changes sign when $\cos^2\theta = 1/d$. As a result, every particle makes a loop, as shown in Figure 1.9. Note the striking difference between the particle trajectories and instantaneous streamlines (see also Exercise 1.6).[11] The permanently displaced mass enclosed between the broken lines (or swept out by any material surface spanning the fluid domain and lying across the direction of motion) is in fact the induced mass itself (Darwin 1953).

1.4 Viscosity

Let us summarize the previous section: neglecting tangential forces (i.e. internal friction) we were able to describe the inertial reaction of the fluid to the body acceleration (quantified by the induced mass). For a motion with a constant speed, we failed to find any force, including the force perpendicular to **u** called lift. If that were true, flying would be impossible. Physical intuition suggests that the resistance force opposite to **u** called drag must be given by the amount of momentum transferred to the fluid in front of the body per unit time (Newton 1687). Multiplying the momentum density ρu by the volume, which is the area R^2 times the velocity u we obtain:

$$F = CR^2 \rho u^2, \tag{1.47}$$

where C is some order-of-unity dimensionless constant (called the drag coefficient) depending on the body shape. This is the correct estimate for the resistance force in the limit of vanishing internal friction. To get a feeling for it, estimate that riding a bike with the speed $4\,\mathrm{m/s} = 14.4\,\mathrm{km/h}$ a body of cross-section $0.75\,\mathrm{m}^2$ meets air drag approximately $12\,N$. A bagel without cream cheese has about $200\,kCal$, which takes a ride of some $7\,\mathrm{km}$ to burn.[12]

Even though (1.47) does not contain viscosity, I don't know any other way to show its validity but to introduce viscosity first and then consider the limit when it vanishes. That limit is quite nontrivial: even an arbitrary small friction makes an infinite region of the flow (called the wake) very much different from the potential flow described in the previous section. Introducing viscosity and describing the wake will take this section and the next one.

1.4.1 Reversibility paradox

Let us consider the absence of resistance in a more general way. We have made five assumptions about the flow: that it is incompressible, irrotational, inviscid (ideal), infinite and steady. The last can always be approached with sufficient precision by waiting long enough (after the body has passed a distance equivalent to a few times its size is usually enough). An irrotational flow of an incompressible fluid is completely determined by the instantaneous body position and velocity. When the body moves with a constant velocity, the flow pattern moves along without changing its form; neither quasimomentum nor kinetic energy of the fluid change so there are no work-doing forces acting between the fluid and the body. If the fluid is finite, that is has a surface, a finite drag arises due to surface waves. If the surface is far away from the body, that drag is negligible.

Could it be that an account of compressibility gives a finite drag for a steady flow, say, due to sound waves carrying energy away? This is not the case, as follows from the *reversibility* of the continuity and Euler equations: the reverse of the flow [defined as $\mathbf{w}(\mathbf{r},t) = -\mathbf{v}(\mathbf{r},-t)$] is also a solution with the velocity at infinity $\mathbf{u}$ instead of $-\mathbf{u}$ but with the same pressure and density fields. For the steady flow, defined by the boundary problem

$$\operatorname{div}\rho\mathbf{v}=0\,,\quad v_n=0 \text{ (on the body surface)}\,,\quad \mathbf{v}\to-\mathbf{u} \text{ at infinity,}$$

$$\frac{v^2}{2}+\int\frac{\mathrm{d}p}{\rho(p)}=\text{const.}\,,$$

the reverse flow $\mathbf{w}(\mathbf{r})=-\mathbf{v}(\mathbf{r})$ has the same pressure field, so it must give the same drag force on the body. Since the drag is supposed to change sign when the direction of motion is reversed, the drag is zero in an ideal irrotational flow. For the particular case of a body with a symmetry, reversibility gives d'Alembert's paradox. For example, if there is a central symmetry, then the pressure on the symmetrical surface elements is the same and the resulting force is a pure couple.[13]

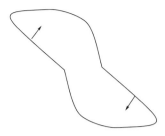

Reversibility paradox teaches us that drag can only come from a force, which is odd with respect to velocity. That requires going beyond an ideal fluid where the pressure $p \propto v^2$ is even. Such drag force can come only from friction which owes its existence to molecular motion in the fluid. To find our way out of the paradoxes of ideal flows toward a real world requires considering internal friction, that is viscosity. Below we shall show that friction provides for drag and lift acting on a body moving through the fluid.

1.4.2 Viscous stress tensor

To describe normal and tangential forces we need now to specify both the orientation of the force component and the orientation of the surface on which it acts. It requires a tensor of the second rank. We define the stress tensor σ_{ij} with the ij element equal to the i component of the force acting on a unit area perpendicular to the j direction. Without a flow, we have only pressure providing the diagonal components, which are normal stresses equal to each other by Pascal's law. Nonuniform flow causes internal friction which changes the stress tensor: $\sigma_{ik} = -p\delta_{ik} + \sigma'_{ik}$ (here the stress is applied to the fluid element under consideration so that the pressure is negative). This changes the momentum flux, $\Pi_{ik} = p\delta_{ik} - \sigma'_{ik} + \rho v_i v_k$, as well as the Euler equation: $\partial \rho v_i / \partial t = -\partial \Pi_{ik} / \partial x_k$.

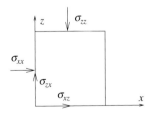

Figure 1.10 Diagonal and nondiagonal components of the stress tensor.

To avoid infinite rotational accelerations, the stress tensor must be symmetric: $\sigma_{ij} = \sigma_{ji}$. Indeed, consider the moment of force (with respect to the axis at the upper right corner) acting on an infinitesimal element with the sizes $\delta x, \delta y, \delta z$:

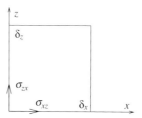

If the stress tensor is not symmetric, then the moment of force $(\sigma_{xz} - \sigma_{zx})$ $\delta x \delta y \delta z$ is nonzero. That moment then must be equal to the time derivative of the angular momentum, which is the moment of inertia $\rho \delta x \delta y \delta z \left[(\delta x)^2 + (\delta z)^2 \right]$ times the angular velocity Ω:

$$(\sigma_{xz} - \sigma_{zx}) \, \delta x \delta y \delta z = \rho \, \delta x \delta y \delta z \left[(\delta x)^2 + (\delta z)^2 \right] \frac{\partial \Omega}{\partial t}.$$

We see that to avoid $\partial \Omega / \partial t \to \infty$ as $(\delta x)^2 + (\delta z)^2 \to 0$ we must assume that $\sigma_{xz} = \sigma_{zx}$.

To connect the frictional stress tensor σ' and the velocity $\mathbf{v}(\mathbf{r})$, note that $\sigma' = 0$ for a uniform flow, so σ' must depend on the velocity spatial derivatives. This simple statement deserves reflection. Molecular motion is expected to provide the flux of any quantity (like heat, concentration of pollutant etc.) determined by the gradient of the quantity. Yet the frictional part of the momentum flux due to molecular motion must be determined by the gradient of velocity rather than the gradient of momentum. Uniformly moving fluid with a nonuniform density does not experience internal friction. That is we assume the medium to be in thermal equilibrium so that there are no fluxes without an inhomogeneous flow.

When the derivatives $\partial v_i / \partial x_k$ are small compared with the velocity changes on a molecular level, one may *assume* that the stress tensor is linearly proportional to the tensor of velocity derivatives (Newton 1687). Fluids with this property are called *Newtonian*. Non-Newtonian fluids are those of elaborate molecular structure (e.g. with long molecular chains, like polymers), where the relation may be non-linear already for moderate strains, and rubber-like liquids, where the stress depends on the fluid's history. For Newtonian fluids, to relate linearly two second-rank tensors, σ'_{ij} and $\partial v_k / \partial x_l$, one generally needs a tensor of the fourth rank: $\sigma'_{ij} = \eta_{ijkl} \partial v_k / \partial x_l$. Yet another simplification comes from the fact that vorticity (that is the antisymmetric part of $\partial v_i / \partial x_j$)

gives no contribution, since it corresponds to a solid-body rotation where no sliding of fluid layers occurs. We thus need to connect two symmetric tensors, the stress σ'_{ij} and the rate of strain $S_{ij} = (\partial v_k/\partial x_l + \partial v_l/\partial x_k)/2$. In the isotropic medium, the viscosity tensor contains only delta-symbols [14]: $\eta_{ijkl} = \eta(\delta_{ik}\delta_{jl} + \delta_{il}\delta_{jk}) + \mu\delta_{ij}\delta_{kl}$. In other words, isotropy requires the principal axes of σ'_{ij} to coincide with those of S_{ij} so that just two constants, η and μ, are left out of the scary fourth-rank tensor:

$$\sigma'_{ij} = \eta(\partial v_i/\partial x_j + \partial v_j/\partial x_i) + \mu\delta_{ij}\partial v_l/\partial x_l. \qquad (1.48)$$

Note that in a nonuniform flow, viscosity not only provides tangential components of the stress tensor but also changes the diagonal components.

Dimensionally $[\eta] = [\mu] = \mathrm{g\,cm^{-1}\,s^{-1}}$. To establish the sign of η, consider a simple shear flow (shown in the figure). The vertical flux of the horizontal momentum is $-\sigma_{xz} = -\eta\,dv_x/dz$, which must be negative, which requires $\eta > 0$.

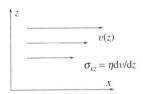

1.4.3 Navier–Stokes equation

Now we substitute σ' into the Euler equation

$$\rho\left(\frac{\partial v_i}{\partial t} + v_k\frac{\partial v_i}{\partial x_k}\right) = -\frac{\partial}{\partial x_k}\left[p\delta_{ik} - \eta\left(\frac{\partial v_i}{\partial x_k} + \frac{\partial v_k}{\partial x_i}\right) - \mu\delta_{ik}\frac{\partial v_l}{\partial x_l}\right]. \qquad (1.49)$$

The viscosity is determined by the thermodynamic state of the system, that is by p, ρ. When p, ρ depend on coordinates, so must $\eta(p,\rho)$ and $\mu(p,\rho)$. However, we consistently assume that the variations of p, ρ are small and put η, μ constant. In this way we get the famous Navier–Stokes equation [15]:

$$\rho\frac{d\mathbf{v}}{dt} = -\nabla p + \eta\Delta\mathbf{v} + (\eta + \mu)\nabla\operatorname{div}\mathbf{v}. \qquad (1.50)$$

Apart from the case of rarefied gases we cannot derive this equation consistently from kinetics. This means only that we generally cannot quantitatively relate η and μ to the properties of the material, the form of the equation is beyond doubt. One can *estimate* the viscosity of a gas, saying that the flux of molecules with thermal velocity v_T through the plane (perpendicular to the velocity gradient) is nv_T and that the molecules come from a layer comparable to the mean free path l and have velocity difference $l\nabla u$, which

causes net momentum flux $mnv_T l\nabla u \simeq \eta\nabla u$, where m is the molecular mass
(again, even if there is a concentration gradient, we do not take it into account).
Therefore, $\eta \simeq mnv_T l = \rho v_T l$. Apparently, this estimate makes sense only
when the velocity changes on the scale far exceeding the mean free path.[16]
The mean free path can be expressed as $l = 1/n\sigma$, where σ is the scattering
cross-section. The viscosity of gases, $\eta \simeq mv_T/\sigma$, is then independent of
density and pressure at a fixed temperature. The thermal velocity grows with
the temperature so that viscosity increases with temperature for gases at
constant density. The scattering cross-section is determined by the strength
of interaction between molecules: the stronger the interaction, the larger is σ,
the shorter is l and the smaller is the viscosity of the gas. Strongly interacting
quark-gluon plasma (in colliders and right after the big bang) is an almost
ideal fluid. We also define kinematic viscosity $\nu = \eta/\rho$, which is estimated as
$\nu \simeq v_T l$.

The estimate $\nu \simeq v_T l$ does not make much sense for a liquid, where a shear
stress is provided by intermolecular forces rather than by molecules coming
from place to place. Stronger interaction then means larger viscosity for
liquids. As temperature increases, molecules move faster and interact weaker,
decreasing the stress, so that ν generally decreases with temperature for liquids
(as any cook will readily confirm). Cooling liquid down to the freezing point,
one increases viscosity to infinity.

Comparing most common liquid and gas, one finds that at room temperature
air has $\nu = 0.15\,\mathrm{cm}^2\,\mathrm{s}^{-1}$ and is kinematically 15 times more viscous than water,
which has $\nu = 0.01\,\mathrm{cm}^2\,\mathrm{s}^{-1}$. That means that if one creates a localized vortex,
it diffuses its vorticity 15 times faster in the air than in the water (see also
Problems 1.9 and 1.15).

The Navier–Stokes equation has higher-order spatial derivatives (second-
order) than the Euler equation so that we need more boundary conditions. Since
we accounted (in the first nonvanishing approximation) for the forces between
fluid layers, we also have to account for the forces of molecular attraction
between a viscous fluid and a solid body surface. Such a force makes the layer
of adjacent fluid to stick to the surface: $\mathbf{v} = 0$ on the surface (not just $v_n = 0$ as
for the Euler equation).[17] When liquid is in contact with a vacuum or rarefied
gas, the boundary cannot support the viscous stress and the boundary condition
is no-stress: $\partial v_l/\partial x_n = 0$. The solutions of the Euler equation generally satisfy
neither no-slip nor no-stress boundary condition. This means that even a very
small viscosity must play a role near a surface.

Viscosity adds an extra term to the momentum flux, but (1.49) and (1.50)
still have the form of a continuity equation that conserves total momentum.
However, viscous friction between fluid layers necessarily leads to some

energy dissipation. Consider, for instance, a viscous incompressible fluid with div $\mathbf{v} = 0$ and calculate the time derivative of the energy at a point:

$$\frac{\rho}{2}\frac{\partial v^2}{\partial t} = -\rho\mathbf{v}\cdot(\mathbf{v}\nabla)\mathbf{v} - \mathbf{v}\cdot\nabla p + v_i\frac{\partial \sigma'_{ik}}{\partial x_k}$$

$$= -\mathrm{div}\left[\rho\mathbf{v}\left(\frac{v^2}{2} + \frac{p}{\rho}\right) - (\mathbf{v}\cdot\sigma')\right] - \sigma'_{ik}\frac{\partial v_i}{\partial x_k}. \tag{1.51}$$

The presence of viscosity results in the momentum flux σ', which is accompanied by the energy transfer, $\mathbf{v}\cdot\sigma'$, and the energy dissipation described by the last term. Because of this last term, this equation does not have the form of a continuity equation and the total energy integral is not conserved. Indeed, after the integration over the whole volume,

$$\frac{dE}{dt} = -\int \sigma'_{ik}\frac{\partial v_i}{\partial x_k}\,dV = -\frac{\eta}{2}\int\left(\frac{\partial v_i}{\partial x_j} + \frac{\partial v_j}{\partial x_i}\right)^2 dV$$

$$= -\eta\int \omega^2 dV < 0. \tag{1.52}$$

The last equality here follows from $\omega^2 = (\varepsilon_{ijk}\partial_j v_k)^2 = (\partial_j v_k)^2 - \partial_k(v_j\partial_j v_k)$, which is true by virtue of $\varepsilon_{ijk}\varepsilon_{ilm} = \delta_{jl}\delta_{km} - \delta_{jm}\delta_{kl}$ and $\partial_i v_i = 0$. Since $\eta > 0$ then viscosity indeed dissipates energy. We see that the dissipation is related to vorticity in an incompressible flow, since the Laplacian of the potential velocity is zero.

The Navier–Stokes equation is a nonlinear partial differential equation of the second order. Not many steady solutions are known. It is particularly easy to find solutions for the geometry where $(\mathbf{v}\cdot\nabla)\mathbf{v} = 0$ and the equation is effectively linear. In particular, symmetry may prescribe that the velocity does not change along itself. One example is the flow along an inclined plane as a model for a river.

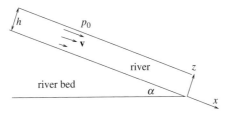

Everything depends only on z. The stationary Navier–Stokes equation takes the form

$$-\nabla p + \eta\Delta\mathbf{v} + \rho\mathbf{g} = 0$$

with z and x projections, respectively,

$$\frac{dp}{dz} + \rho g \cos \alpha = 0,$$

$$\eta \frac{d^2 v}{dz^2} + \rho g \sin \alpha = 0. \qquad (1.53)$$

The boundary condition on the bottom is $v(0) = 0$. On the surface, the boundary condition is that the stress should be normal and balance the pressure: $\sigma_{xz}(h) = \eta \, dv(h)/dz = 0$ and $\sigma_{zz}(h) = -p(h) = -p_0$. The solution is simple:

$$p(z) = p_0 + \rho g (h - z) \cos \alpha, \qquad v(z) = \frac{\rho g \sin \alpha}{2\eta} z(2h - z). \qquad (1.54)$$

Let us see how it corresponds to reality. Take water with the kinematic viscosity $\nu = \eta/\rho = 10^{-2}\,\mathrm{cm}^2\,\mathrm{s}^{-1}$. For a rain puddle with thickness $h = 1$ mm on a slope $\alpha \sim 10^{-2}$ we get a reasonable estimate $v \sim 5\,\mathrm{cm\,s}^{-1}$. For slow plain rivers (like the Nile or the Volga) with $h \simeq 10$ m and $\alpha \simeq 0.3\,\mathrm{km}/3000\,\mathrm{km} \simeq 10^{-4}$ one gets $v(h) \simeq 100\,\mathrm{km\,s}^{-1}$ which is evidently impossible (the resolution of this dramatic discrepancy is that real rivers are turbulent, as discussed in Section 2.2.3). What distinguishes puddle and river, why are they not similar? To answer this question, we need to characterize flows by a dimensionless parameter.

1.4.4 Law of similarity

One can obtain some important conclusions about flows from a dimensional analysis. Consider a steady incompressible flow past a body described by the equation

$$(\mathbf{v} \cdot \nabla)\mathbf{v} = -\nabla p/\rho + \nu \Delta \mathbf{v}$$

and by the boundary conditions $\mathbf{v}(\infty) = \mathbf{u}$ and $\mathbf{v} = 0$ on the surface of a body of size L. For a given body shape, both $\mathbf{v}$ and p/ρ are functions of coordinates $\mathbf{r}$ and three variables, $\mathbf{u}, \nu, L$. Out of the latter, one can form only one dimensionless quantity, called the Reynolds number

$$Re = uL/\nu. \qquad (1.55)$$

This is the most important parameter in this book, since it determines the ratio of the nonlinear (inertial) term $(\mathbf{v} \cdot \nabla)\mathbf{v}$ to the viscous friction term $\nu \Delta \mathbf{v}$. Indeed, in the dimensionless variables $\mathbf{v}/u$, $\mathbf{r}/L$, tu/L the incompressible Navier–Stokes equation takes the form

$$\frac{\partial \mathbf{v}}{\partial t} + (\mathbf{v} \cdot \nabla)\mathbf{v} = -\frac{\nabla p}{\rho} + \frac{1}{Re} \Delta \mathbf{v}. \qquad (1.56)$$

Since the kinematic viscosity is the thermal velocity times the mean free path, the Reynolds number is

$$Re = \frac{u}{v_{\mathrm{T}}} \frac{L}{l} \ .$$

We see that within the hydrodynamic limit ($L \gg l$), Re can be both large and small depending on the ratio $u/v_{\mathrm{T}} \simeq u/c$.

Dimensionless velocity must be a function of dimensionless variables: $\mathbf{v} = u\mathbf{f}(\mathbf{r}/L, Re)$ – it is a unit-free relation. Flows that correspond to the same Re can be obtained from one another simply by changing the units of v and r; such flows are called similar (Reynolds 1883). In the same way, $p/\rho = u^2\varphi(\mathbf{r}/L, Re)$. For a quantity independent of coordinates, only some function of Re is unknown – the drag or lift force, for instance, must be $F = \rho u^2 L^2 f(Re)$. This law of similarity is exploited in modeling: to measure, say, the drag on a ship that one is designing, one can build a smaller model yet pull it faster through the fluid (or use a less viscous fluid).

The Reynolds number, as a ratio of inertia to friction, makes sense for all types of flow as long as u is some characteristic velocity and L is a scale of the velocity change. For the inclined plane flow (1.54), the nonlinear term (and the Reynolds number) is zero since $\mathbf{v} \perp \nabla\mathbf{v}$. How much does one need to perturb this alignment to make $Re \simeq 1$? Such perturbations always exist in reality where the bottom is never perfectly flat. One may think that bottom imperfections are more important for a shallow paddle than for a deep river. It is the other way around. Denoting $\pi/2 - \beta$ the angle between $\mathbf{v}$ and $\nabla\mathbf{v}$ we get $Re(\beta) = v(h)h\beta/v \simeq g\alpha\beta h^3/v^2$. For a puddle, $Re(\beta) \simeq 50\beta$ while for a river $Re(\beta) \simeq 10^{12}\beta$. It is then clear that the (so-called laminar) solution (1.54) may make sense for a puddle, but for a river it must be distorted by even tiny bottom imperfections, see Figure 1.11.

Gravity brings another dimensionless parameter, the Froude number $Fr = u^2/Lg$; the flows are similar for the same Re and Fr. Such parameters (changes of which bring qualitative changes in the regime even for fixed geometry and boundary conditions) are called control parameters.[18]

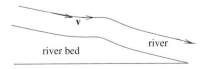

Figure 1.11 The nonflat bottom of the river bed makes the velocity of the river change along itself, which leads to a nonzero inertial term $(\mathbf{v} \cdot \nabla)\mathbf{v}$ in the Navier–Stokes equation.

The law of similarity is a particular case of the so-called π-theorem: Assume that among all m variables $\{b_1, \ldots, b_m\}$ we have only $k \leq m$ dimensionally independent quantities – this means that the dimensionalities $[b_{k+1}], \ldots, [b_m]$ could be expressed via $[b_1], \ldots, [b_k]$ like $[b_{k+j}] = \prod_{l=1}^{l=k} [b_l]^{\beta_{jl}}$. Then all dimensionless quantities can be expressed in terms of $m - k$ dimensionless variables $\pi_1 = b_{k+1}/\prod_{l=1}^{l=k} b_l^{\beta_{1l}}, \ldots, \pi_{m-k} = b_m/\prod_{l=1}^{l=k} b_l^{\beta_{m-k,l}}$. For example, the three above quantities u, v, L have two independent dimensionalities, cm and sec, which allows one to introduce the single dimensionless parameter, the Reynolds number.

1.5 Stokes flow and the wake

We now return to the flow past a body armed with the knowledge of internal friction. Unfortunately, the Navier–Stokes equation is a nonlinear partial differential equation which we cannot solve in a closed analytical form even for a flow around a sphere. We shall therefore proceed in the way that physicists often do: solve a limiting case of very small Reynolds numbers and then use this solution to understand higher-Re flow. Remember that in Section 1.3 we failed spectacularly to describe high-Re flow as an ideal fluid. This time we shall realize, with the help of qualitative arguments and experimental data, that when viscosity becomes very small its effect stays finite. On the way we shall learn new notions of a boundary layer and a separation phenomenon. Our reward will be the resolution of paradoxes and the formulae for drag and lift.

1.5.1 Slow motion

Consider such a slow motion of a body through the fluid that the Reynolds number, $Re = uR/v$, is small. This means that we can neglect inertia. Indeed, if we stop pushing the body, friction stops it after a time of order R^2/v, so that inertia moves it by the distance of order $uR^2/v = R \cdot Re$, which is much less than the body size R. Formally, neglecting inertia means omitting the nonlinear term $(\mathbf{v} \cdot \nabla)\mathbf{v}$ in the Navier–Stokes equation. This makes our problem linear so that the fluid velocity is proportional to the body velocity: $v \propto u$. The viscous stress (1.48) and the pressure are also linear in u and so must be the drag force:

$$F = \int \sigma \mathrm{d}f \simeq \int \mathrm{d}f \eta u/R \simeq 4\pi R^2 \eta u/R = 4\pi \eta u R.$$

This crude estimate coincides with the true answer given later by (1.60) up to the dimensionless factor $3/2$. Linear proportionality between the force and the velocity makes the low-Reynolds flow an Aristotelean world.

Now, if you wish to know what force would move a body with $Re \simeq 1$ (or $1/6\pi$ for a sphere), you find amazingly that such a force, $F \sim \eta^2/\rho$, does not depend on the body size (that is, it is the same for a bacterium and a ship). For water, $\eta^2/\rho \simeq 10^{-4}\,\mathrm{dyn} = 10^{-9}\,\mathrm{N}$.

If we also assume than no fast-changing forces is applied to the fluid then the whole inertia term, $\rho\, dv/dt = \rho[\partial v/\partial t + (v\nabla)v]$, can be neglected:

$$\partial v/\partial t \simeq (v\nabla)v \simeq u^2/L \ll v\Delta v \simeq vu/L^2.$$

In this case, the Navier–Stokes equation (1.50) turns into the Stokes equation:

$$\eta \Delta \mathbf{v} = \nabla p . \tag{1.57}$$

One can also say that this equation describes the flow of a massless fluid, sometimes called creeping flow. In particular, motion on microscopic and nanoscopic scales in fluids usually corresponds to very low Reynolds numbers and is described by the Stokes equation. Swimming at low Re is very different from pushing water backwards as we do at finite Re. One defines swimming as changing shape in a periodic way to move. First, there is no inertia at low Re so that momentum diffuses instantly through the fluid. Therefore, it does not matter how quickly or slowly we change the shape. What matters is the shape changes itself, i.e. low-Re swimming is purely geometrical. Second, linearity means that simply retracing the changes back (by inverting the forces, i.e. the pressure gradients) we just retrace the motion. One thus needs to change a shape periodically but in a time-irreversible way, that is to have a cycle in a configuration space. Microorganisms do that by sending progressive waves along their surfaces. Every point of a surface may move time-reversibly (even in straight lines); the time direction is encoded in the phase shift between different points. For example, a spermatozoid swims by sending helical waves down its tail.[19] See Exercises 1.10 and 3.11 for other examples.

Probably the simplest way to find solutions of the Stokes equation (1.57) is to reduce it again to the Laplace equation. Since creeping flows practically always can be treated as incompressible, we can use $div\,\mathbf{v} = 0$ to apply the operator div to (1.57) and obtain $\Delta p = 0$. Let us solve this equation for the flow around a sphere when the only vector in the problem is the velocity $\mathbf{u}$. The respective solution of the Laplace equation for a scalar (this time it is pressure rather than potential) is again a dipole:

$$p - p_0 = \frac{c(\mathbf{u} \cdot \mathbf{n})}{r^2} .$$

Indeed, positive and negative pressure variations must have the same magnitude on respective points of the sphere. We now differentiate it and substitute into (1.57). We can obtain the partial solution of the resulting equation with the radial velocity perturbation decaying as an inverse distance: $\mathbf{v} - \mathbf{u} = c\mathbf{n}(\mathbf{u} \cdot \mathbf{n})/2\eta r$. To make it incompressible we need to add a solution of the homogeneous equation: $\mathbf{v} - \mathbf{u} = c[\mathbf{u} + \mathbf{n}(\mathbf{u} \cdot \mathbf{n})]/2\eta r$. This solution, however, cannot satisfy the boundary condition on the sphere surface. For that we need to add yet another solution of the Laplace equation that is the potential part:

$$\mathbf{v} = \mathbf{u} + c\frac{\mathbf{u} + \mathbf{n}(\mathbf{u} \cdot \mathbf{n})}{2\eta r} + b\frac{3\mathbf{n}(\mathbf{u} \cdot \mathbf{n}) - \mathbf{u}}{r^3} . \qquad (1.58)$$

The boundary condition $\mathbf{v}(R) = 0$ gives $\mathbf{u}$ component $1 + c/2\eta R - b/R^3 = 0$ and $\mathbf{n}$ component $c/2\eta R + 3b/R^3 = 0$ so that $c = -6\eta R/4$ and $b = R^3/4$. In spherical components

$$v_r = u\cos\theta \left(1 - \frac{3R}{2r} + \frac{R^3}{2r^3} \right),$$
$$v_\theta = -u\sin\theta \left(1 - \frac{3R}{4r} - \frac{R^3}{4r^3} \right). \qquad (1.59)$$

Sanity check confirms that $c < 0$, that is the pressure is larger upstream so that fluid flows down the pressure gradient. The vorticity too satisfies the Laplace equation and is a dipole field:

$$\Delta\,\mathrm{curl}\,\mathbf{v} = \Delta\omega = 0 \Rightarrow \omega = c'\frac{[\mathbf{u} \times \mathbf{n}]}{r^2}$$

with $c' = -3R/2$ from $\nabla p = \eta\Delta\mathbf{v} = -\eta\,\mathrm{curl}\,\omega$.

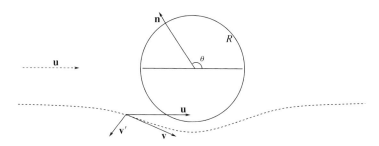

Stokes' formula for the drag. The force acting on a unit surface is the momentum flux through it. On a solid surface $\mathbf{v} = 0$ and $F_i = -\sigma_{ik}n_k = pn_i - \sigma'_{ik}n_k$. In our case, the only nonzero component is along $\mathbf{u}$:

$$F_x = \int (-p\cos\theta + \sigma'_{rr}\cos\theta - \sigma'_{r\theta}\sin\theta)\,\mathrm{d}f$$
$$= (3\eta u/2R)\int \mathrm{d}f = 6\pi R\eta u . \qquad (1.60)$$

Here, we substituted $\sigma'_{rr} = 2\eta\,\partial v_r/\partial r = 0$ at $r = R$ and

$$p(R) = -\frac{3\eta u}{2R}\cos\theta,$$

$$\sigma'_{r\theta}(R) = \eta\left(\frac{1}{r}\frac{\partial v_r}{\partial\theta} + \frac{\partial v_\theta}{\partial r} - \frac{v_\theta}{r}\right) = -\frac{3\eta u}{2R}\sin\theta.$$

The viscous force is tangential while the pressure force is normal to the surface. The vertical components of the forces cancel each other at every point – the sphere pushes fluid strictly forward and the force is purely horizontal. The viscous and pressure contributions sum into the horizontal force $3\eta u/2R$, which is independent of θ, i.e. the same for all points on the sphere. The force is determined by the dynamic viscosity and independent of the fluid density. The density will enter via the dimensionless Reynolds number $Re = \rho uR/\eta$ when we account for fluid inertia. The formula (1.60) is called Stokes's law; it works well until $Re \simeq 0.5$.

We can now generalize the equation of motion for particles in a flow, (1.35), by including viscous force. Considering for simplicity a heavy spherical particle with the radius a and neglecting fluid inertia, we relate the particle velocity $\mathbf{u}$ to the fluid velocity $\mathbf{v}$:

$$\frac{d\mathbf{u}}{dt} = \frac{\mathbf{v} - \mathbf{u}}{\tau}, \qquad \tau = \frac{2\rho_0 a^2}{9\rho\nu}. \tag{1.61}$$

Formal solution of this equation,

$$\mathbf{u}(t) = \int_{-\infty}^{t} e^{(t'-t)/\tau}\mathbf{v}(t')\,dt', \tag{1.62}$$

is expressed via the fluid velocity in the reference frame co-moving with the particle. The solution shows that the particle motion is retarded by the response time τ which is called Stokes time. For millimeter-sized rain droplets $\tau \simeq 10\,\mathrm{sec}$, while for micron-size cloud droplets it is million times less.

The solution of the boundary value problem for the Laplace equation is unique. So if we require the pressure gradient to go to zero at infinity, then a source of any form having constant pressure on the surface produces constant pressure in the whole space. Similarly, since the vorticity must tend to zero at infinity, it can be nonzero in space only if it is nonzero on the source surface. In particular, a point source produces a radial flow having a constant pressure and zero vorticity (see Exercise 1.16). By superposition this is also true for an arbitrary combination of point sources and sinks since creeping flows are linear. In two dimensions, pressure and vorticity can be treated as, respectively, real and imaginary part of an analytic function, which brings power of complex analysis to the problem. In particular, identically zero pressure gradient means zero vorticity and vice versa.

1.5.2 The boundary layer and the separation phenomenon

Another thing named after Stokes is the paradox – no finite solution of (1.57) exists for flow past a body in two dimensions. Indeed, the dipole fields in two dimensions must decay as $1/r$. When pressure and vorticity decay as $1/r$, the velocity logarithmically *grows* with r (rather than saturates as in 3d). Stokes himself believed that moving a long cylinder through a very viscous fluid one continually "increases the quantity of fluid which it carries with it." Rayleigh later pointed out that for the Stokes flow the viscous term decays with the distance faster than the inertial term, so that the latter must be taken into account sufficiently far from the body. Indeed, the assumption of small Reynolds number requires in any dimensionality

$$v\nabla v \simeq u^2 R/r^2 \ll v\Delta v \simeq vuR^2/r^3 , \qquad (1.63)$$

so that the Stokes equation is valid for $r \ll v/u$. One can call v/u the width of the viscous boundary layer. The existence of the boundary layer resolves Stokes paradox: inertia must stop the growth of velocity perturbation at $r \simeq v/u$.

If we put a cylinder in a uniform flow having velocity u at infinity, then $v(r = v/u) \simeq u$, and the steady solution at $R \leq r < v/u$ is as follows:

$$v(r) \simeq u \frac{\log(r/R)}{\log(v/uR)}. \qquad (1.64)$$

That helps to understand another 2D peculiarity: the drag on a cylinder is a force per unit length f, for which the only dimensionally possible combination linear in viscosity is ρuv, absurdly independent of the size R. We expect the drag to increase with R (and turn to zero as $R \to 0$), so that the dimensionless drag $f/\rho uv$ must go to zero when $Re = uR/v \to 0$. Indeed, the friction force is determined by the fluid velocity gradient on the cylinder, which according to (1.64) depends logarithmically on the size: $f \simeq \rho uv/\log Re^{-1}$, that is the drag is logarithmically suppressed at small Re. Nonlinear dependence of the force on the viscosity in two dimensions shows that no matter how small is Re, one cannot neglect inertia and consider fluid massless as long as any fluid element is infinite in the third direction.

In both two and three dimensions the Stokes flows are realized inside the boundary layer under the assumption that the size of the body is much less than the width of the layer. So what is the flow outside the viscous boundary layer, that is for $r > v/u$? Is it potential? The answer is "yes" only for very small *Re*. For finite *Re*, there is an infinite region (called the *wake*) behind the body where it is impossible to neglect viscosity whatever the distance from the body. This is because viscosity produces vorticity in the boundary layer:

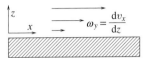

At small *Re*, the process that dominates the flow is vorticity diffusion away from the body caused by friction. The Stokes approximation, $\omega \propto [\mathbf{u} \times \mathbf{n}]/r^2$, corresponds to symmetrical diffusion of vorticity in all directions. In particular, the flow has a left–right (fore-and-aft) symmetry. For finite *Re*, vorticity production by friction is accompanied by inertial vorticity advection; it is then intuitively clear that the flow upstream and downstream from the body must be different since the body leaves vorticity behind it. Indeed, when the boundary layer is comparable to the body size or less, the fluid particles enter the layer with zero vorticity but leave it with a nonzero vorticity, which is then carried further since inertia dominates friction outside. Therefore, there should exist some downstream region reached by fluid particles which move along streamlines passing through the boundary layer. The flow in this region (wake) is essentially rotational. On the other hand, streamlines that do not pass through the boundary layer correspond to almost potential motion.

Let us describe qualitatively how the wake arises. The phenomenon called *separation* is responsible for wake creation (Prandtl 1905). Consider, for instance, the flow around a cylinder, shown in Figure 1.12. The ideal fluid flow is symmetrical with respect to the plane AB. The point D is a stagnation point. On the upstream half DA, the fluid particles accelerate and the pressure decreases according to the Bernoulli theorem. On the downstream part AC, the reverse happens, that is every particle moves against the pressure gradient. Let us add viscosity to the picture. A small viscosity changes pressure only slightly across the boundary layer. Indeed, if the viscosity is small, the boundary layer is thin and can be considered locally flat. Denote *u* the velocity right outside the

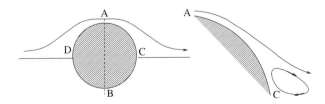

Figure 1.12 Symmetric streamlines for an ideal flow (left) and appearance of separation and a recirculating vortex in a viscous fluid (right).

boundary layer. In the boundary layer, at $z < v/u$, no-slip condition prescribes $v_x \simeq u^2 z/v$ and $\partial v_x/\partial x \simeq u^2 z/vR \simeq \partial v_z/\partial z$. The normal velocity is then $v_z \simeq u^2 z^2/vR$, which gives the pressure gradient, $\partial p/\partial z = -\rho(v\nabla)v - \eta\Delta v_z \simeq \rho u^2/R$, so that the pressure change across the layer is $\rho u^2/Re$ that is small when Re is large. In other words, the pressure inside the boundary layer is almost equal to that in the main stream, which is the pressure of the ideal fluid flow. But the velocities of the fluid particles that reach the points A and B are lower in a viscous fluid than in an ideal fluid because of viscous friction in the boundary layer. Then those particles have insufficient energy to overcome the pressure gradient downstream. The particle motion in the boundary layer is stopped by the pressure gradient before the point C is reached. The pressure gradient then becomes the force that accelerates the particles from the point C upwards, producing separation [20] and a recirculating vortex. A similar mechanism is responsible for recirculating eddies in the corners [21] shown at the end of Section 1.2.4.

Reversing the flow pattern of separation one obtains attachment: jets tend to attach to walls and merge with each other. Consider first a jet in an infinite fluid and denote the velocity along the jet u. The momentum flux through any section is the same: $\int u^2\, df = \text{const}$. On the other hand, the energy flux, $\int u^3\, df$, decreases along the jet owing to viscous friction. This means that the mass flux of the fluid, $\int u\, df$, must grow – a phenomenon known as *entrainment*.[22] When the jet has a wall (or another jet) on one side, it draws into itself less fluid from this side and so inclines toward the wall until it is attached, as shown in the figure. The jet then can stay attached and follow the surface even when it is convex.

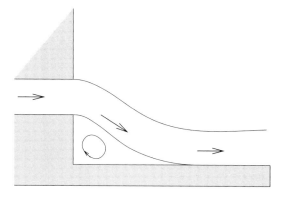

wall-attaching jet

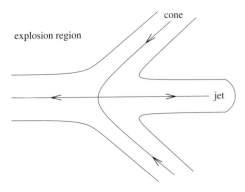

Figure 1.13 Scheme of the flow of a cumulative jet in the reference frame moving with the cone.

In particular, jet merging explains a cumulative effect of armour-piercing shells, which contain a conical void covered by a metal and surrounded by explosives. Explosion turns the metal into a fluid. Moreover, the pressure is so high that the tangential stresses can be neglected and the fluid flow is ideal, see Exercise 1.18. Fluid moves toward the axis where it creates a cumulative jet with a high momentum density (Lavrent'ev 1947, Taylor 1948), see Figure 1.13 and Exercise 1.15. Similarly, if one creates a void in a liquid with, say, a raindrop or other falling object then the vertical momentum of the liquid that rushes to fill the void creates a jet, shown in Figure 1.14.

1.5.3 Flow transformations

Let us now use the case of the flow past a cylinder to describe briefly how the flow pattern changes as the Reynolds number goes from small to large. The flow is most symmetric for $Re \ll 1$ when it is steady and has an exact up–down symmetry and approximate (order Re) left–right symmetry. Separation of the boundary layer and the occurrence of eddies is a change of the flow topology; it occurs around $Re \simeq 5$. The first loss of exact symmetries happens around $Re \simeq 40$ when the flow becomes periodic in time. This happens because the recirculating eddies don't have enough time to spread; they are being detached from the body and carried away by the flow as new eddies are generated. Periodic flow with shedding eddies has up–down and continuous time shift symmetries broken and replaced by a combined symmetry of up–down reflection and time shift for half a period. The shedding of eddies explains many surprising symmetry-breaking phenomena, like, for instance,

Figure 1.14 Jet shooting out after the droplet fall. Upper image – beginning of the jet formation; lower image – jet formed. Photograph copyright: Sdtr, Rmarmion, www.dreamstime.com.

an air bubble rising through water (or champagne) in a zigzag or spiral rather than a straight path.[23] It also has a strong effect on swimming and flying at moderate Re; many fish, birds and hovering insects are able to exploit the pressure variation resulted from vortex formation and detachment. For flow past a body, vortex shedding results in a double train of vortices, called the Kármán vortex street[24] behind the body, as shown in Figure 1.15. Kármán vortex street is responsible for many acoustic phenomena like the roar of propeller, sound caused by a wind rushing past a tree or the swish of a whip.

As the Reynolds number increases further, the vortices become unstable and produce an irregular turbulent motion downstream, as seen in Figure 1.16.[25] That turbulence is three-dimensional, i.e. the translational invariance along the cylinder is broken as well. The higher Re, the closer to the body turbulence starts. At $Re \simeq 10^5$, the turbulence reaches the body, making the rear part of the

Figure 1.15 Kármán vortex street behind a cylinder at $Re = 105$. Photograph by Sadatoshi Taneda, reproduced from *J. Phys. Soc. Japan*, **20**, 1714 (1965).

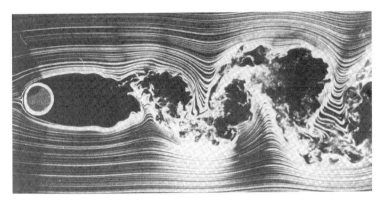

Figure 1.16 Flow past a cylinder at $Re = 10^4$. Photograph by Thomas Corke and Hassan Najib, reproduced from [27].

boundary layer turbulent and bringing the so-called drag crisis (discovered by Eiffel in 1912 and explained by Prandtl in 1914): Since a turbulent boundary layer entrains more fluid from outside, has thus more momentum and separates later downstream than a laminar one, the wake area gets smaller and the drag is reduced. One can check that for $Re < 10^5$ a stick encounters more drag when moving through a still fluid than when kept still in a moving fluid (in the latter case the flow is usually turbulent before the stick so that the boundary layer is turbulent as well). Generations of scientists, starting from Leonardo Da Vinci, believed that the drag must be the same (despite experience telling otherwise) because of Galilean invariance, which, of course, is applicable only to an infinite uniform flow, not to real streams.

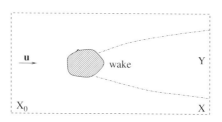

Figure 1.17 Scheme of the wake.

1.5.4 Drag and lift with a wake

We can now describe the way Nature resolves reversibility and d'Alembert's paradoxes. As in Section 1.3, we consider the steady flow far from the body and relate it to the force acting on the body. The new experimental wisdom we now have is the existence of the wake (Figure 1.17). The flow is irrotational outside the boundary layer and the wake. It is convenient to relate the force to the momentum flux through a closed surface. For a dipole potential flow $v \propto r^{-3}$ from Section 1.3, that flux was zero for a distant surface. Now the wake gives a finite contribution. The total momentum flux transported by the fluid through any closed surface is equal to the rate of momentum change, which is equal to the force acting on the body:

$$F_i = \oint \Pi_{ik} \mathrm{d} f_k = \oint (p_0 + p') \delta_{ik} + \rho (u_i + v_i)(u_k + v_k) \, \mathrm{d} f_k \qquad (1.65)$$

$$= (p_0 \delta_{ik} + \rho u_i u_k) \oint \mathrm{d} f_k + \rho u_i \oint v_k \, \mathrm{d} f_k + \oint [p' + \rho (u_k v_i + v_i v_k)] \, \mathrm{d} f_k.$$

In the last line, the first integral vanishes because the surface is closed and the second one because of mass conservation: $\rho \oint v_k \, \mathrm{d} f_k = 0$. Far from the body $v \ll u$ and we neglect terms quadratic in v:

$$F_i \approx \left(\iint_{X_0} - \iint_{X} \right) (p' \delta_{ix} + \rho u v_i) \, \mathrm{d} y \mathrm{d} z. \qquad (1.66)$$

Drag with a wake. Consider the x component of the force (1.66):

$$F_x = \left(\iint_{X_0} - \iint_{X} \right) (p' + \rho u v_x) \, \mathrm{d} y \mathrm{d} z.$$

Outside the wake we have potential flow where the Bernoulli relation, $p + \rho |\mathbf{u} + \mathbf{v}|^2 / 2 = p_0 + \rho u^2 / 2$, gives $p' \approx -\rho u v_x$ so that the integral outside the wake vanishes. Inside the wake, the pressure is about the same (since it does not change across the almost straight streamlines, as we argued in Section 1.5.2) but the velocity perturbation v_x is shown below to be much larger than outside, so that

$$F_x = -\rho u \int\!\!\int_{\text{wake}} v_x \, \mathrm{d}y \mathrm{d}z. \tag{1.67}$$

Force is positive (directed to the right) since v_x is negative. The integral in (1.67) is equal to the deficit of fluid flux Q through the wake area (i.e. the difference between the flux with and without the body). That deficit is x-independent, which has dramatic consequences for the potential flow outside the wake because it has to compensate for the deficit. That means that the integral $\int \mathbf{v} \, \mathrm{d}\mathbf{f}$ outside the wake is also r-independent which requires $v \propto r^{-2}$. This corresponds to the potential flow with the source equal to the flow deficit: $\phi = Q/r$. It is analogous to a charge field in electrostatics. We had thrown away this source flow in Section 1.3 but now we see that it exceeds the dipole flow $\phi = \mathbf{A} \cdot \nabla(1/r)$ (which we had without the wake) and dominates sufficiently far from the body.

The wake breaks the fore-and-aft symmetry and thus resolves the paradoxes, providing for a nonzero drag in the limit of vanishing viscosity. Now that we learnt that drag is related to the vorticity production by the body, we can appreciate how slender fish swim by passing a wave along its body. As explained in Section 1.2.4, small oscillations produce little vorticity, so the fish keeps the wave amplitude small over most of the body, increasing it toward the tail.

It is important that the wake has an infinite length under stationary conditions, otherwise the body and the finite wake could be treated as a single entity and we are back to paradoxes. The behavior of the drag coefficient $C(Re) = F/\rho u^2 R^2$ is shown in Figure 1.18. Notice the drag crisis, which gives

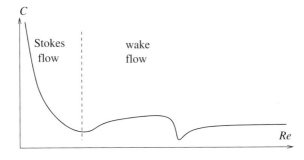

Figure 1.18 Sketch of the drag dependence on the Reynolds number.

the lowest C. To understand why $C \to$ const. as $Re \to \infty$ and prove (1.47), one needs to go a long way, developing the theory of turbulence briefly described in the next chapter.

The lift is the force component of (1.66) perpendicular to **u**:

$$F_y = \rho u \left(\int_{X_0} - \int_X \right) v_y \mathrm{d}y \mathrm{d}z. \qquad (1.68)$$

It is also determined by the wake – without the wake the flow is potential with $v_y = \partial \phi / \partial y$ and $v_z = \partial \phi / \partial z$ so that $\int v_y \mathrm{d}y \mathrm{d}z = \int v_z \mathrm{d}y \mathrm{d}z = 0$ since the potential is zero at infinity. We have seen in (1.34) that purely potential flow produces no lift. Without friction-caused separation, birds and planes would not be able to fly. For a wing, which is long in the z-direction, the lift force per unit length can be related to the velocity circulation around the wing. Indeed, adding and subtracting (vanishing) integrals of v_x over two $y = \pm$ const. lines we turn (1.68) into

$$F_y = \rho u \oint \mathbf{v} \cdot \mathrm{d}\mathbf{l}. \qquad (1.69)$$

Circulation over the contour is equal to the vorticity flux through the contour, which is again due to the wake.

To fly effectively, wings must minimize the drag proportional to the wake area, keeping the lift proportional to the circulation. This is achieved by making wings slender with the thickness h much less than the width l and by making the leading edge smooth and the trailing edge sharp. One can often hear a simple explanation of the lift of the wing as being the result of $v_2 > v_1 \Rightarrow P_2 < P_1$. This is basically true and does not contradict the above argument.

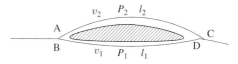

The point is that the circulation over the closed contour ACDB is non-zero: $v_2 l_2 > v_1 l_1$. Indeed, the Bernoulli relation, $P_2 - P_1 = \rho(v_1^2 - v_2^2)/2 = \rho(v_1 - v_2)(v_1 + v_2)/2 \approx \rho(v_1 - v_2)u$, gives an alternative derivation of the lift via circulation: $\int (P_2 - P_1)\mathrm{d}x \approx \rho u \int (v_1 - v_2)\mathrm{d}x$. It would be wrong, however, to argue that $v_2 > v_1$ because $l_2 > l_1$ – neighbouring fluid elements A,B do not meet again at the trailing edge; C is shifted forward relative to D. To understand that, first, note that the velocity difference has also horizontal component and is proportional to the wing thickness h, or in other words, to the small parameter h/l. However, the difference in the path lengths can be estimated as

$l_2 - l_1 \simeq h^2/l$. Therefore, for a slender wing, the upper fluid element reaches the trailing edge before the lower one. Nonzero circulation around the body in translational motion requires a wake. For a slender wing with a sharp trailing edge, the wake is very thin, like a cut, and a nonzero circulation means a jump of the potential ϕ across the wake.One can generalize the method of complex potential from Section 1.2.4 for describing flows with circulation, which involves logarithmic terms.[26]

Note that to have lift, one needs to break the up–down symmetry. Momentum conservation suggests that one can also relate the lift to the downward deflection of the flow by the body. One can also look at at the pressure, which decreases toward the center of curvature of streamlines. If there is a body, wing or sail, which curves the streamlines, then the pressure decreases away from its inner part toward the center of curvature and increases away from its outer part; since the pressure far away from the body is the same on either side, then the pressure is lower on the outer side and the force is directed away from the center of curvature.

One can have a nonzero circulation and deflection of a flow without any wake simply by rotating the moving body. Since there is a nonzero circulation, then there is a deflecting (Magnus) force acting on a rotating moving sphere (Figure 1.19). That force is well known to all ball players, from soccer to tennis. The air travels faster relative to the centre of the ball where the ball surface is moving in the same direction as the air. This reduces the pressure, while on the other side of the ball the pressure increases. The result is a lift force, perpendicular to the motion. (As J. J. Thomson put it, "The ball follows its nose.") One can roughly estimate the magnitude of the Magnus force from the pressure difference between the two sides, which is proportional to the translation velocity u times the rotation frequency Ω:

$$\Delta p \simeq \rho[(u+\Omega R)^2 - (u-\Omega R)^2]/2 = 2\rho u\Omega R . \qquad (1.70)$$

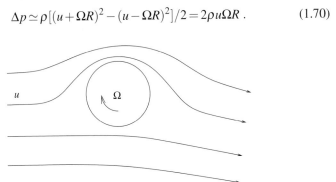

Figure 1.19 Streamlines around a rotating body.

Newton argued that a rotating ball curves because the side that moves faster meets more resistance. Since he considered the resistance force proportional to the velocity squared that is to the pressure, this gives the same estimate (1.70). The Magnus force is exploited by winged seeds, which travel away from the parent tree superimposing rotation on their descent[27]; it also acts on quantum vortices moving in superfluids or superconductors (see Exercise 1.11).

Note that the drag force changes sign upon time reversal, when $\mathbf{v} \to -\mathbf{v}$, while the lift force and the Magnus force don't.

Moral: wake existence teaches us that small viscosity changes the flow not only in the boundary layer but also in the whole space, both inside and outside the wake. Physically, this is because vorticity is produced in the boundary layer and is transported outside into the wake. Vorticity diffuses in low-Reynolds flows and concentrates in wakes at high Reynolds number. On the other hand, even for a very large viscosity, inertia dominates sufficiently far from the body.[28]

It is instructive to think about similarities and differences in the ways that vorticity penetrating the bulk makes life interesting in classical fluids versus quantum fluids and superconductors. An evident difference is that vorticity is continuous in a classical fluid while vortices are quantized in quantum fluids. The quant of circulation $\hbar/m$ has the dimensionality of viscosity. It is also instructive to compare the Navier–Stokes equation $\mathbf{v}_t = \nu \Delta \mathbf{v} + \ldots$ with the Schrödinger equation, $\psi_t = \mathrm{i}(\hbar/m)\Delta\psi + \ldots$, where i makes the Laplacian term non-dissipative. Similarity is that both ν and $\hbar/m$ are singular perturbations that introduce the highest spatial derivative and change the boundary conditions, leaving anomalies when they go to $+0$.

Exercises

1.1 Proceeding from the fact that the force exerted across any plane surface is wholly normal, prove that its intensity (per unit area) is the same for all aspects of the plane (Pascal's law).

1.2 Consider a self-gravitating fluid with the gravitational potential ϕ related to the density by

$$\Delta\phi = 4\pi G\rho,$$

G being the constant of gravitation. Assume spherical symmetry and static equilibrium. Describe the radial distribution of pressure for an incompressible liquid and an isothermal ideal gas.

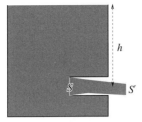

Figure 1.20 Borda mouthpiece.

1.3 Find the discharge rate from a small orifice with a cylindrical tube, projecting inward (Figure 1.20). Assume h, S and the acceleration due to gravity, g given. Does such a hole correspond to the limiting (smallest or largest) value of the "coefficient of contraction" S'/S? Here S is the orifice area and S' is the area of the jet where contraction ceases (vena contracta).

1.4 Prove that if you put a little solid particle – not an infinitesimal point – at any place in the liquid it will rotate with angular velocity Ω equal to half of the local vorticity $\omega = \operatorname{curl} \vec{v}$: $\Omega = \omega/2$.

1.5 There is a permanent source of water at the bottom of a large reservoir. Find the maximal elevation of the water surface for two cases:

(i) a straight narrow slit with the constant influx q ($\mathrm{g\,cm^{-1}\,s^{-1}}$) per unit length;

(ii) a point-like source with the influx Q ($\mathrm{g\,s^{-1}}$).

The fluid density is ρ, the depth of the fluid far away from the source is h. The acceleration due to gravity is g. Assume that the flow is potential.

1.6 Sketch streamlines for the potential inviscid flow and for the viscous Stokes flow in two reference systems, in which: (i) the fluid at infinity is at rest; (ii) the sphere is at rest.

1.7 A heavy ball with density ρ_0 is connected to a spring and has oscillation frequency ω_a. The same ball attached to a rope makes a pendulum with oscillation frequency ω_b. How do those frequencies change if such oscillators are placed in an ideal fluid with density ρ? What change is brought about by a small viscosity of the fluid ($\nu \ll \omega_{a,b} a^2$ where a is the ball radius and ν is the kinematic viscosity)?

1.8 An underwater explosion released energy E and produced a gas bubble oscillating with period T, which is known to be completely determined by E, the static pressure p in the water and the water density ρ. Find the form of the dependence $T(E,p,\rho)$ up to a numerical factor. If the initial radius a is known instead of E, can we determine the form of the dependence $T(a,p,\rho)$?

1.9   At $t=0$ a straight vortex line exists in a viscous fluid. In cylindrical coordinates, it is described as follows: $v_r=v_z=0$, $v_\theta=\Gamma/2\pi r$, where Γ is some constant. Find the vorticity $\omega(r,t)$ as a function of time and the time behavior of the total vorticity $\int \omega(r)r\,dr$.

1.10 To appreciate how one swims in a syrup, consider the so-called Purcell swimmer shown in Figure 1.21. It can change its shape by changing separately the angles between the middle link and the arms. Assume that the angle θ is small. The numbers correspond to consecutive shapes. In position 5 the swimmer has the same shape as in position 1 but moved in space. Which direction? What distinguishes this direction? How does the displacement depend on θ?

1.11 In making a free kick, a good soccer player is able to utilize the Magnus force to send the ball around the wall of defenders. Neglecting vertical motion, estimate the horizontal deflection of the ball (with radius $R=11$ cm and weight $m=450$ g, according to FIFA rules) sent with side spin 10 revolutions per second and speed $v_0=30$ m s^{-1} toward the goal, which is $L=30$ m away. Take air density ρ to be 10^{-3} g cm^{-3}.

1.12 Like flying, sailing also utilizes the lift (perpendicular) force acting on the sails and the keel. The fact that the wind provides a force perpendicular to the sail allows one even to move against the wind. But most optimal for starting and reaching maximal speed, as all windsurfers know, is to orient the board perpendicular to the wind and set the sail at about 45 degrees, as in Figure 1.22. Why? Draw the forces acting on the board. Does the board move exactly in the direction at which the keel is pointed? Can one move faster than the wind?

1.13 Consider spherical water droplet falling under gravity.

(i) Find the fall velocity in air. Droplet radius is 0.01 mm. Air and water viscosities and densities are, respectively, $\eta_a=1.8\cdot10^{-4}$ g s^{-1} cm^{-1}, $\eta_w=0.01$ g s^{-1} cm^{-1} and $\rho_a=1.2\cdot10^{-3}$ g cm^{-3}, $\rho_w=1$ g cm^{-3};

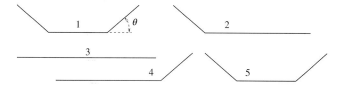

Figure 1.21 Subsequent shapes of the swimmer.

Figure 1.22 Left panel: the sailor holds the sail against the wind which is thus coming from behind her back. Right panel: scheme of the position of the board and its sail with respect to the wind. Photograph copyright: Paul Topp, www.dreamstime.com.

(ii) Describe the motion of an initially small droplet falling in a saturated cloud and absorbing the vapour in a swept volume so that its volume grows proportionally to its velocity and its cross-section. Consider a quasisteady approximation, when the droplet acceleration is much less than the acceleration due to gravity, g.

1.14 Consider planar free jets in an ideal fluid in the geometry shown in the figure. Find how the widths of the outgoing jets depend on the angle $2\theta_0$ between the impinging jets.

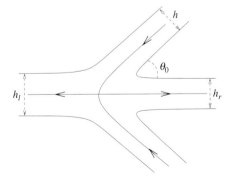

1.15 Can viscosity stop fluid rotation?

1.16 Consider spherically symmetric flow with radial velocity $v_r \propto r^{1-d}$ where d is space dimensionality. Compute the vorticity, the viscous stress tensor and the viscous force. Is the flow dissipative? Does it need a pressure gradient to sustain it?

1.17 Gas flows steadily through a long pipe under the action of the pressure gradient. The pipe walls are rough so that molecules are reflected from the walls isotropically forgetting the incoming velocity. Does the mass flow rate increase or decrease with the gas density at fixed temperature? Hint: consider limits of low/high density when the pipe radius is smaller/larger than the mean free path of collisions between molecules.

1.18 An underground charge explodes at the depth h far exceeding the charge size r_0. Most soils consist of densely packed rigid (quartz) grains which are weakly cemented. Assume that the explosion produces pressure large enough to break tangential stresses between grains and make soil flowing. When tangential stresses can be neglected, pressure-dominated flow can be considered ideal and potential. On the other hand, assume the pressure small enough so it cannot deform the grains and the flow can be considered incompressible. That usually works for pressures between hundreds and tens of thousands of bars (1 bar$= 10^5$ N m^{-2}). The region of crushed material is bounded by the surface at which the material velocity becomes equal some critical velocity c. Find the radius of the crater on the surface if the energy impacted into the fluid motion is E and the soil density is ρ.

2

Unsteady flows

Fluid flows can be kept steady only for very low Reynolds numbers and for velocities much less than the velocity of sound. Otherwise, either flow experiences instability and becomes turbulent or sound and shock waves are excited. Both sets of phenomena are described in this chapter.

A formal reason for instability is nonlinearity of the equations of fluid mechanics. For incompressible flows, the only nonlinearity is due to fluid inertia. We shall see how a perturbation of a steady flow can grow due to inertia, thus causing an instability. For large Reynolds numbers, the development of instabilities leads to a strongly fluctuating state of turbulence.

An account of compressibility, on the other hand, leads to another type of unsteady phenomena: sound waves. When density perturbation is small, velocity perturbation is much less than the speed of sound and the waves can be treated within the framework of perturbation theory. We first consider linear acoustics and discover what phenomena appear as long as one accounts for a finiteness of the speed of sound. We then consider nonlinear acoustic phenomena, the creation of shocks and acoustic turbulence.

2.1 Instabilities

At large Re most of the steady solutions of the Navier–Stokes equation are unstable and generate an unsteady flow called turbulence.

2.1.1 Kelvin–Helmholtz instability

Apart from a uniform flow in the whole space, the simplest steady flow of an ideal fluid is a uniform flow in a semiinfinite domain with the velocity parallel to the boundary. Physically, it corresponds to one fluid layer sliding

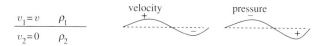

Figure 2.1 Tangential velocity discontinuity (left) and the physics of Kelvin–Helmholtz instability (right).

along another. Mathematically, it is a tangential velocity discontinuity, which is a formal steady solution of the Euler equation. It is a crude approximation of the description of wakes and shear flows. This simple solution is unstable with respect to arguably the simplest instability, as described by Helmholtz (1868) and Kelvin (1871). The dynamics of the Kelvin–Helmholtz instability is easy to see from Figure 2.1 where $+$ and $-$ denote, respectively, increase and decrease in velocity and pressure brought by surface modulation. The part of the boundary that moved toward the still region has the velocity lower, and the pressure higher, than over the part that moved toward the flow. Such pressure distribution further increases the modulation of the surface.

The perturbations $\mathbf{v}'$ and p' satisfy the following system of equations

$$\operatorname{div} \mathbf{v}' = 0, \qquad \frac{\partial \mathbf{v}'}{\partial t} + v \frac{\partial \mathbf{v}'}{\partial x} = -\frac{\nabla p'}{\rho}.$$

Applying the divergence operator to the second equation we get $\Delta p' = 0$. This means that the elementary perturbations have the following form

$$p'_1 = \exp[\mathrm{i}(kx - \Omega t) - kz],$$
$$v'_{1z} = -\mathrm{i}k p'_1 / \rho_1 (kv - \Omega).$$

Indeed, the solutions of the Laplace equation that are periodic in one direction must be exponential in another direction.

The solution on the other side is obtained by setting $v = 0$ and $z \to -z$. Eigenvalue Ω is to be found from matching the solutions at the boundary. To relate the upper side (indexed 1) to the lower side (indexed 2) we introduce $\zeta(x,t)$, the elevation of the surface, its time derivative is the z component of the velocity:

$$\frac{\mathrm{d}\zeta}{\mathrm{d}t} = \frac{\partial \zeta}{\partial t} + v \frac{\partial \zeta}{\partial x} = v'_z, \qquad (2.1)$$

that is $v'_z = \mathrm{i}\zeta(kv - \Omega)$ and $p'_1 = -\zeta \rho_1 (kv - \Omega)^2 / k$. On the other side, we can express in a similar way $p'_2 = \zeta \rho_2 \Omega^2 / k$. The pressure is continuous across the surface, which gives the matching condition:

$$\rho_1 (kv - \Omega)^2 = -\rho_2 \Omega^2 \quad \Rightarrow \quad \Omega = kv \frac{\rho_1 \pm \mathrm{i}\sqrt{\rho_1 \rho_2}}{\rho_1 + \rho_2}. \qquad (2.2)$$

Positive ImΩ means an exponential growth of perturbations, i.e. instability.[1] The largest growth rate corresponds to the largest admissible wavenumber. In reality the transitional layer, where velocity increases from zero to v, has some finite thickness δ and our approach is valid only for $k\delta \ll 1$.

A complementary insight into the physics of the Kelvin–Helmholtz instability can be obtained by considering vorticity. In the unperturbed flow, the vorticity $\partial v_x / \partial z$ is concentrated in the transitional layer, which is thus called the vortex layer (or *vortex sheet* when $\delta \to 0$). One can consider a discrete version of the vortex layer as a chain of identical vortices, shown in Figure 2.2. Owing to symmetry, such an infinite array of vortex lines is stationary since the velocities imparted to any given vortex by all the others cancel. The small displacements shown by straight arrows in Figure 2.2 lead to an instability with the vortex chain breaking into pairs of vortices circling round one another. That circling motion turns an initially sinusoidal perturbation into spiral rolls during the nonlinear stage of the evolution, as shown in Figure 2.3, obtained experimentally. The Kelvin–Helmholtz instability in the atmosphere is often made visible by corrugated cloud patterns, as seen in Figure 2.4; similar patterns are seen on sand dunes. It is also believed to be partially responsible for clear air turbulence (that is atmospheric turbulence unrelated to moist convection). Numerous manifestations of this instability are found in astrophysics, from the interface between the solar wind and the Earth's magnetosphere to the boundaries of galactic jets.

Figure 2.2 The array of vortex lines is unstable with respect to the displacements, shown by straight arrows.

Figure 2.3 Spiral vortices generated by the Kelvin–Helmholtz instability. Photograph by F. Roberts, P. Dimotakis and A. Roshko, reproduced from [27].

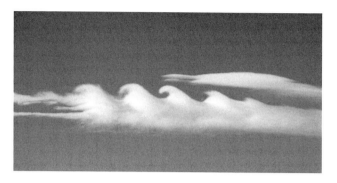

Figure 2.4 Lower cloud shows the pattern of breaking waves generated by the Kelvin–Helmholtz instability. Photo authorship and copyright: Brooks Martner.

The vortex view of the Kelvin–Helmholtz instability suggests that a uni-directional flow depending on a single transverse coordinate, like $v_x(z)$, can only be unstable if it has a vorticity maximum on some surface. Such a vorticity maximum is an inflection point of the velocity since $d\omega/dx = d^2 v_x/dz^2$. This explains why flows without inflection points are linearly stable (Rayleigh 1880). Examples of such flows are plane linear profiles, flows driven by pressure gradients in a pipe or channel, flows between two planes moving with different velocities, etc.[2] In particular, the vortex layer can be locally considered as a linear profile for perturbations with $k\delta \gg 1$, so such perturbations cannot grow. Therefore, the maximal growth rate corresponds to $k\delta \simeq 1$, i.e. the wavelength of the most unstable perturbation is comparable to the layer thickness.

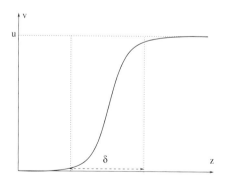

Our consideration of the Kelvin–Helmholtz instability was for completely inviscid fluids which presumes that the effective Reynolds number was large: $Re = v\delta/v \gg 1$. In the opposite limit when the friction is very strong, the velocity profile is not stationary but rather evolves according to the equation $\partial v_x(z,t)/\partial t = v \partial^2 v_x(z,t)/\partial z^2$, which describes the thickness growing as $\delta \propto \sqrt{vt}$. Such a diffusing vortex layer is stable because the friction damps all the perturbations. It is thus clear that there must exist a threshold Reynolds number above which instability is possible. We now consider this threshold from a general energetic perspective.

2.1.2 Energetic estimate of the stability threshold

The energy balance between the unperturbed steady flow $\mathbf{v}_0(\mathbf{r})$ and the perturbation $\mathbf{v}_1(\mathbf{r},t)$ helps one to understand the role of viscosity in imposing an instability threshold. Consider the flow $\mathbf{v}_0(\mathbf{r})$, which is a steady solution of the Navier–Stokes equation $(\mathbf{v}_0 \cdot \nabla)\mathbf{v}_0 = -\nabla p_0/\rho + v\Delta \mathbf{v}_0$. The perturbed flow $\mathbf{v}_0(\mathbf{r}) + \mathbf{v}_1(\mathbf{r},t)$ satisfies the equation:

$$\frac{\partial \mathbf{v}_1}{\partial t} + (\mathbf{v}_1 \cdot \nabla)\mathbf{v}_0 + (\mathbf{v}_0 \cdot \nabla)\mathbf{v}_1 + (\mathbf{v}_1 \cdot \nabla)\mathbf{v}_1$$
$$= -\frac{\nabla p_1}{\rho} + v\Delta \mathbf{v}_1 . \tag{2.3}$$

Making the scalar product of (2.3) with $\mathbf{v}_1$ and using incompressibility one gets:

$$\frac{1}{2}\frac{\partial v_1^2}{\partial t} = -v_{1i}v_{1k}\frac{\partial v_{0i}}{\partial x_k} - \frac{1}{Re}\frac{\partial v_{1i}}{\partial x_k}\frac{\partial v_{1i}}{\partial x_k}$$
$$- \frac{\partial}{\partial x_k}\left[\frac{v_1^2}{2}(v_{0k} + v_{1k}) + p_1 v_{1k} - \frac{v_{1i}}{Re}\frac{\partial v_{1i}}{\partial x_k}\right].$$

The last term disappears after integration over the volume:

$$\frac{\mathrm{d}}{\mathrm{d}t}\int \frac{v_1^2}{2}\mathrm{d}\mathbf{r} = T - \frac{D}{Re}, \tag{2.4}$$
$$T = -\int v_{1i}v_{1k}\frac{\partial v_{0i}}{\partial x_k}\,\mathrm{d}\mathbf{r},$$
$$D = \int \left(\frac{\partial v_{1i}}{\partial x_k}\right)^2 \mathrm{d}\mathbf{r}.$$

The term T is due to inertial forces and the term D is due to viscous friction. We see that for stability (i.e. for decay of the energy of the perturbation) one needs friction to dominate over inertia (Reynolds 1894):

$$Re < Re_E = \min_{v_1} \frac{D}{T} . \tag{2.5}$$

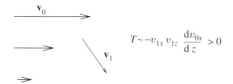

Figure 2.5 Orientation of the perturbation velocity $\mathbf{v}_1$ with respect to the steady shear $\mathbf{v}_0$ that provides for an energy flux from the shear to the perturbation: the flux $v_{1x}v_{1z}$ of the momentum x-component is carried along z from higher to lower values of the mean flow v_{0x}.

The minimum is taken over different perturbation flows. Since both T and D are quadratic in the perturbation velocity, their ratio depends on the orientation and spatial dependence of $\mathbf{v}_1(\mathbf{r})$ but not on its magnitude. Since a uniform flow is stable, then it is natural that one needs $\partial v_0/\partial r \neq 0$ to provide nonzero energy input T. Moreover, one must have the perturbation velocity oriented in such a way as to have both the component v_{1i} along the mean flow and the component v_{1k} along the gradient of the mean flow. Then $v_{1i}v_{1k}$ is the flux of the momentum along the flow in the direction of the flow gradient. To have positive T one needs this flux to have the sign opposite to the flow gradient; in this case, the perturbation diminishes the flow gradient and the flow energy, thus increasing the perturbation energy. An example of such a geometry is shown in Figure 2.5. While the flow is always stable for $Re < Re_E$, it is not necessary unstable when one can find a perturbation that breaks (2.5); for instability to develop, the perturbation must also evolve in such a way as to keep $T > D$. As a consequence, the critical Reynolds numbers are usually somewhat higher than those given by the energetic estimate.

2.1.3 Landau's law

Dimensionless parameter, such as Reynolds number, whose change can bring qualitative changes in behavior, is called control parameter. When the control parameter passes a critical value, the system undergoes an instability and goes into a new state. Generally, one cannot say much about this new state except for the case when it is not very much different from the old one. That may happen when the control parameter is not far from critical. Consider $Re > Re_{cr}$ but $Re - Re_{cr} \ll Re_{cr}$. Just above the instability threshold, there is usually only one unstable mode. Let us linearize the equation (2.3) with respect to the perturbation $\mathbf{v}_1(\mathbf{r},t)$, i.e. omit the term $(\mathbf{v_1} \cdot \nabla)\mathbf{v_1}$. The resulting linear differential equation with time-independent coefficients has a solution of the form $\mathbf{v}_1 = \mathbf{f}_1(\mathbf{r})\exp(\gamma_1 t - i\omega_1 t)$ which describes the unstable mode. The exponential growth has to be restricted by the terms that are nonlinear in $\mathbf{v}_1$. The solution of a weakly nonlinear equation can be sought in the form $\mathbf{v}_1 = \mathbf{f}_1(\mathbf{r})A(t)$. The equation for the amplitude $A(t)$ must generally have the following form: $d|A|^2/dt = 2\gamma_1|A|^2 +$ third-order terms $+\cdots$. The fourth-order terms are obtained by expanding $\mathbf{v} = \mathbf{v}_0 + \mathbf{v}_1 + \mathbf{v}_2$ further and accounting for $\mathbf{v}_2 \propto \mathbf{v}_1^2$ in the equation for $\mathbf{v}_1$. The growth rate turns into zero at $Re = Re_{cr}$ and generally $\gamma_1 \propto Re - Re_{cr}$, while the frequency is usually finite at $Re \to Re_{cr}$. We can thus average the amplitude equation over time larger than $2\pi/\omega_1$ but smaller than $1/\gamma_1$. Since the time of averaging contains many periods, then among the terms of the third and fourth order only $|A|^4$ gives a nonzero contribution:

$$\overline{\frac{d|A|^2}{dt}} = 2\gamma_1|A|^2 - \alpha|A|^4. \tag{2.6}$$

Since the averaging time is much less than the time of the modulus change, then one can remove the overbar in the left-hand side of (2.6) and solve it as a usual ordinary differential equation. This equation has the solution

$$|A|^{-2} = \alpha/2\gamma_1 + \text{const.} \cdot \exp(-2\gamma_1 t) \to \alpha/2\gamma_1.$$

The saturated value changes with the control parameter according to the so-called Landau's law:

$$|A|^2_{max} = \frac{2\gamma}{\alpha} \propto Re - Re_{cr}.$$

We thus see that nonlinearity (i.e. inertia) has a dual role: it overcomes friction to make the instability possible but then it stops the growth of the perturbation at a finite amplitude. If $\alpha < 0$ then one needs a $-\beta|A|^6$ term in (2.6) to stabilize the instability

$$\overline{\frac{d|A|^2}{dt}} = 2\gamma_1|A|^2 - \alpha|A|^4 - \beta|A|^6. \tag{2.7}$$

The saturated value is now

$$|A|^2_{max} = -\frac{\alpha}{2\beta} \pm \sqrt{\frac{\alpha^2}{4\beta^2} + \frac{2\gamma_1}{\beta}}.$$

Stability with respect to the variation of $|A|^2$ within the framework of (2.7) is determined by the factor $2\gamma_1 - 2\alpha|A|^2_{max} - 3\beta|A|^4_{max}$. Between B and C, the steady flow is metastable. The broken curve is unstable.

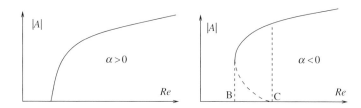

This description is based on the assumption that at $Re - Re_{cr} \ll Re_{cr}$ the only important dependence is $\gamma_1(Re)$ very much like in Landau's theory of phase transitions (which also treats loss of stability). The amplitude A, which is nonzero on one side of the transition, is an analogue of the order parameter. Cases of positive and negative α correspond to the phase transitions of the second and first order, respectively.

2.2 Turbulence

As the Reynolds number increases beyond the threshold of the first instability, it eventually reaches a value where the new periodic flow becomes unstable in its own turn with respect to another type of perturbation, usually with smaller scale and consequently higher frequency. Every new instability brings about an extra degree of freedom, characterized by the amplitude and the phase of the new periodic motion. The phases are determined by (usually uncontrolled) initial perturbations. At very large Re, a sequence of instabilities produces *turbulence* as a superposition of motions of different scales (Figure 2.6). The resulting flow is irregular both spatially and temporally so we need to describe it statistically. In this Section we describe how turbulence flows appear, what are their basic properties and how they carry and mix substances.

2.2.1 Turbulence onset

Flows that undergo several instabilities can be effectively characterized by a finite number of (generally interacting) degrees of freedom, which could be described by the complex amplitudes of the excited modes $A_i(t)$. Such flows become temporally chaotic already at moderate Re because motion in the phase space of more than three interacting degrees of freedom may tend asymptotically to sets (called attractors), which are more complicated than fixed points (steady states), cycles (periodic motions) or tori (multiperiodic motions). Namely, there exist chaotic attractors that consist of saddle-point trajectories. Such trajectories have stable directions, by which the system approaches the

Figure 2.6 Instabilities in three almost identical convective jets lead to completely different flow patterns. Notice also the appearance of progressively smaller scales as the instabilities develop. Photograph copyright: Vbotond, www.dreamstime.com.

attractor, and unstable directions, lying within the attractor. Because all trajectories are unstable on the attractor, any two initially close trajectories separate exponentially at a mean rate called the Lyapunov exponent. To intuitively appreciate how the mean stretching rate can be positive in a random flow, note that around a saddle-point more vectors undergo stretching than contraction (Exercise 2.1). Exponential separation of trajectories means instability and unpredictability of the flow patterns. The resulting fluid flow that corresponds to a chaotic attractor in phase space is regular in space and random in time; it is called dynamical chaos.[3] One can estimate the Lyapunov exponent for the Earth's atmosphere by dividing the typical wind velocity $20\,\mathrm{m\,s^{-1}}$ by the global scale $10\,000\,\mathrm{km}$. The inverse Lyapunov exponent gives the time one can reasonably hope to predict weather, which is $10^7\,\mathrm{m}/(20\,\mathrm{m\,s^{-1}}) = 5 \cdot 10^5\,\mathrm{s}$, i.e. about a week. One can also study Lagrangian chaos and related mixing due to exponential separation of the trajectories of fluid particles (rather than trajectories in the phase space), this will be described in Section 2.2.4 below.

The laminar flow can be linearly stable at large Re (as unidirectional flows without inflection points) or at all Re. However, the basin of attraction of such a flow shrinks when Re grows, so that fluctuations of small yet finite amplitude are able to excite turbulence, which then sustains itself. Turbulence onset is of probabilistic nature in this case: for any finite Re, there is a finite probability for the flow to return to a laminar state. For example, some fraction of the pipe flow is always laminar. This is because there is always a finite probability for any given perturbation either to decay or to expand/split. The mean decay time increases while the mean split time decreases with Re; one can define the critical Re as when the times are equal.[4] As the Reynolds number increases, expansion and splitting of perturbations leads to filling the space with turbulence. On the contrary, linearly unstable laminar flow cannot exist beyond the critical value of the Reynolds number or other control parameter. In such systems, turbulence arises from an increase in temporal complexity of fluid motion.

2.2.2 Cascade

Here we consider turbulence at very large Re, which is random in space and in time. Such flows require a statistical description that is able to predict mean (expectation) values of different quantities. Despite five centuries of effort (since Leonardo da Vinci) a complete description is still lacking but some important elements have been established. We do not yet know exactly which properties of developed turbulence are independent of the path taken to $Re \to \infty$, in particular, whether it started from a linearly unstable laminar flow and went through stochastic attractors or needed finite perturbations of a linearly stable flow. However, a revealing insight into the universal aspects of turbulence is given a cascade picture (Figure 2.7), which I present in this section. It is a useful phenomenology, both from a fundamental viewpoint of understanding a state with many degrees of freedom deviated from equilibrium and from a practical viewpoint of explaining the empirical fact that the drag force is finite in the inviscid limit. The finiteness of the drag coefficient, $C(Re) = F/\rho u^2 L^2 \to$ const. at $Re \to \infty$ (see Figure 1.18), means that the rate of the kinetic energy input per unit mass, $\varepsilon = Fu/\rho L^3 = Cu^3/2L$, stays finite when $v \to 0$. Where does all this energy go if we consider not an infinite wake but a bounded flow, say, generated by a permanently acting fan in a room? Experiment (and everyday experience) tells us that a fan generates some air flow whose magnitude stabilizes after a while, which means that the input is balanced by the viscous dissipation. Since the input is expected to be independent of small viscosity, this means that the energy dissipation rate $\varepsilon = v \int \omega^2 \, dV/V$ stays finite when $v \to 0$ (if the fluid temperature is kept constant).

Historically, the understanding of turbulence as cascade started from an empirical law established by Richardson (observing seeds and balloons released in the wind): the mean squared distance between two particles in turbulence increases in a super-diffusive way, $\langle R^2(t) \rangle \propto t^3$. Here the average is over different pairs of particles. The parameter that can relate $\langle R^2(t) \rangle$ and t^3 must have dimensionality $cm^2 s^{-3}$, which is that of the dissipation rate ε: $\langle R^2(t) \rangle \simeq \varepsilon t^3$. Richardson's law can be *interpreted* as the increase of the

Figure 2.7 Cascade. Photograph copyright: Lee 2010, www.dreamstime.com.

typical velocity difference $\delta v(R)$ with distance R: since there are vortices of different scales in a turbulent flow, the velocity difference at a given distance is due to vortices with comparable scales and smaller; as the distance increases, more (and larger) vortices contribute to the relative velocity, which makes separation faster than diffusive (when the velocity is independent of the distance). Richardson's law suggests the law of the relative velocity increase with the distance in turbulence. Indeed, $R(t) \simeq \varepsilon^{1/2} t^{3/2}$ is a solution of the equation $dR/dt \simeq (\varepsilon R)^{1/3}$; since $dR/dt = \delta v(R)$ then

$$\delta v(R) \simeq (\varepsilon R)^{1/3} \Rightarrow \frac{(\delta v)^3}{R} \simeq \varepsilon. \tag{2.8}$$

The last relation brings the idea of the energy cascade over scales, which goes from the scale L with $\delta v(L) \simeq u$ down to the viscous scale l defined by $l \delta v(l) \simeq v$. The viscous scale is much larger than the mean free path as long as $\delta v(L) \ll c$. The energy flux through the given scale R can be estimated as the energy $(\delta v)^2$ divided by the time $R/\delta v$. For the so-called inertial interval of scales, $L \gg R \gg l$, there is neither force nor dissipation so that the energy flux $\varepsilon(R) = \langle \delta v^3(R) \rangle / R$ may be expected to be R-independent, as suggested by (2.8). When $v \to 0$, the viscous scale l decreases, that is the cascade gets longer, but the amount of the flux and the dissipation rate stay the same. In other words, finiteness of ε in the limit of vanishing viscosity can be interpreted as locality of the energy transfer in R-space (or equivalently, in Fourier space). By using an analogy, one may say that turbulence is supposed to work as a pipe with a flux through its cross-section independent of the length of the pipe.[5]

Let us illustrate the cascade picture by an estimate. For a wind $v \simeq 10 \, \text{m/sec}$ at the height $L \simeq 150 \, \text{m}$, we shall have $Re \simeq 10^8$ and the viscous scale can be estimated from $l \delta v(l) \simeq l v (l/L)^{1/3} \simeq v$, which gives $l = L Re^{-3/4} \simeq 0.15 \, \text{mm}$. So wind turbulence contains vortices from hundreds of meters to the fraction of a millimeter.

The cascade picture is a nice phenomenology but can one support it with any derivation? One can obtain an exact relation that quantifies the flux constancy (Kolmogorov 1941). Let us derive first the equation for the correlation function of the velocity at different points for an idealized turbulence whose statistics are presumed isotropic and homogeneous in space. We assume no external forces so that the turbulence must decay with time. Let us find the time derivative of the correlation function of the components of the velocity difference between the points 1 and 2,

$$\langle (v_{1i} - v_{2i})(v_{1k} - v_{2k}) \rangle = \frac{2\langle v^2 \rangle}{3} \delta_{ik} - 2 \langle v_{2i} v_{1k} \rangle.$$

The time derivative of the kinetic energy is minus the dissipation rate: $\varepsilon = -\mathrm{d}\langle v^2 \rangle / 2\mathrm{d}t$. To obtain the time derivative of the two-point velocity correlation function, take the Navier–Stokes equation at some point $\mathbf{r}_1$, multiply it by the velocity $\mathbf{v}_2$ at another point $\mathbf{r}_2$ and average it over time intervals larger than $|\mathbf{r}_1 - \mathbf{r}_2| / |\mathbf{v}_1 - \mathbf{v}_2|$ and smaller than L/u:

$$\frac{\partial}{\partial t} \langle v_{1i} v_{2k} \rangle = -\frac{\partial}{\partial x_{1l}} \langle v_{1l} v_{1i} v_{2k} \rangle - \frac{\partial}{\partial x_{2l}} \langle v_{1i} v_{2k} v_{2l} \rangle$$
$$- \frac{1}{\rho} \frac{\partial}{\partial x_{1i}} \langle p_1 v_{2k} \rangle - \frac{1}{\rho} \frac{\partial}{\partial x_{2k}} \langle p_2 v_{1i} \rangle + v(\Delta_1 + \Delta_2) \langle v_{1i} v_{2k} \rangle.$$

It is presumed that the temporal average is equivalent to the spatial average, property called ergodicity. Statistical isotropy means that the vector $\langle p_1 \mathbf{v}_2 \rangle$ has nowhere to look but to $\mathbf{r} = \mathbf{r}_1 - \mathbf{r}_2$, the only divergence-less such vector, $\mathbf{r}/r^3$, does not satisfy the finiteness at $r = 0$ so that $\langle p_1 \mathbf{v}_2 \rangle = 0$. Owing to the space homogeneity, all the correlation functions depend only on $\mathbf{r} = \mathbf{r}_1 - \mathbf{r}_2$.

$$\frac{\partial}{\partial t} \langle v_{1i} v_{2k} \rangle = -\frac{\partial}{\partial x_l} \left(\langle v_{1l} v_{1i} v_{2k} \rangle + \langle v_{2i} v_{1k} v_{1l} \rangle \right) + 2v\Delta \langle v_{1i} v_{2k} \rangle. \qquad (2.9)$$

We have used here $\langle v_{1i}v_{2k}v_{2l}\rangle = -\langle v_{2i}v_{1k}v_{1l}\rangle$ since under $1 \leftrightarrow 2$ both $\mathbf{r}$ and a third-rank tensor change sign (the tensor turns into zero when $1 \rightarrow 2$). By straightforward yet lengthy derivation[6] one can rewrite (2.9) for the moments of the longitudinal velocity difference, called structure functions,

$$S_n(r,t) = \langle [\mathbf{r}\cdot(\mathbf{v}_1 - \mathbf{v}_2)]^n /r^n\rangle \ .$$

This gives the so-called Kármán–Howarth relation

$$\frac{\partial S_2}{\partial t} = -\frac{1}{3r^4}\frac{\partial}{\partial r}\left(r^4 S_3\right) - \frac{4\varepsilon}{3} + \frac{2\nu}{r^4}\frac{\partial}{\partial r}\left(r^4 \frac{\partial S_2}{\partial r}\right) \ . \qquad (2.10)$$

The average quantity S_2 changes only together with a large-scale motion, so

$$\frac{\partial S_2}{\partial t} \simeq \frac{S_2 u}{L} \ll \frac{S_3}{r}$$

at $r \ll L$. On the other hand, we consider $r \gg l$, or more formally we consider finite r and take the limit $\nu \rightarrow 0$ so that the last term disappears. We *assume* now that ε has a finite limit at $\nu \rightarrow 0$ and obtain Kolmogorov's 4/5 law:

$$S_3(r) = -4\varepsilon r/5 \ . \qquad (2.11)$$

That remarkable relation tells that turbulence is irreversible since S_3 does not change sign when $t \rightarrow -t$ which requires $\mathbf{v} \rightarrow -\mathbf{v}$. If screen a movie of turbulence backwards, we can indeed tell that something is wrong! This is what is called "anomaly" in modern field-theoretical language: a symmetry of the inviscid equation (here, time-reversal invariance) is broken by the viscous term even though the latter might have been expected to become negligible in the limit $\nu \rightarrow 0$. In other words, the effect of symmetry breaking remains finite when symmetry-breaking factor goes to zero.

It is instructive to recall that the statistics is reversible in thermal equilibrium: the detailed balance principle states that the probabilities of every process and its time reversal are equal. This is related to the fact that the thermostat provides both dissipation and short-correlated forcing, which balance each other at every scale and timescale, as expressed by the fluctuation-dissipation theorem. On the contrary, irreversibility of turbulence statistics can be traced to the fact that forcing and dissipation act on different scales.

Here the good news ends. There is no analytic theory to give us other structure functions. One may *assume* following Kolmogorov (1941) that ε is the only quantity determining the statistics in the inertial interval, then on dimensional grounds $S_n \simeq (\varepsilon r)^{n/3}$. Experiment gives the power laws, $S_n(r) \propto r^{\zeta_n}$ but with the exponents ζ_n deviating from $n/3$ for $n \neq 3$. Moments of the velocity difference can be obtained from the probability density function (PDF), which describes the probability of measuring the velocity difference $\delta v = u$ at distance r: $S_n(r) = \int u^n \mathscr{P}(u,r)\,du$. Deviations of ζ_n

from $n/3$ mean that the PDF $\mathscr{P}(\delta v, r)$ is not scale invariant, i.e. cannot be presented as $(\delta v)^{-1}$ times the dimensionless function of the single variable $\delta v/(\varepsilon r)^{1/3}$. Apparently, there is more to turbulence than just a cascade, and ε is not all one must know to predict the statistics of the velocity. Similar breakdown of scale invariance takes place in the simpler one-dimensional case of Burgers turbulence described in Section 2.3.4 below, where it can be related to shock waves. One can also relate breakdown of scale invariance in turbulence to statistical integrals of motion of fluid particles For example, for two particles with the coordinates $\mathbf{R}_1(t), \mathbf{R}_2(t)$ and velocities $\mathbf{v}_1(t), \mathbf{v}_2(t)$ the quantity $\langle |\mathbf{v}_1 - \mathbf{v}_2|^2 |\mathbf{R}_1 - \mathbf{R}_2|^{-\zeta_2} \rangle$ does not change at $t \to \infty$.[7] Both symmetries, one broken by pumping (scale invariance) and another by friction (time reversibility) are not restored even when $r/L \to 0$ and $l/r \to 0$.

To appreciate difficulties in turbulence theory, one can cast the turbulence problem into that of quantum field theory. Consider the Navier–Stokes equation driven by a random force $\mathbf{f}$ with the Gaussian probability distribution $P(\mathbf{f})$ defined by the variance $\langle f_i(0,0) f_j(\mathbf{r},t) \rangle = D_{ij}(\mathbf{r},t)$. Then the probability of any flow $\mathbf{v}(\mathbf{r},t)$ is given by the Feynman path integral over velocities satisfying the Navier–Stokes equation with different force histories:

$$\int D\mathbf{v} D\mathbf{f} \delta\left(\partial_t \mathbf{v} + (\mathbf{v}\cdot\nabla)\mathbf{v} + \nabla P/\rho - \nu\Delta\mathbf{v} - \mathbf{f}\right) P(\mathbf{f})$$
$$= \int D\mathbf{v} D\mathbf{p} \exp\left[-D_{ij}p_i p_j + \mathrm{i}p_i\left(\partial_t v_i + v_k\nabla_k v_i + \nabla_i P/\rho - \nu\Delta v_i\right)\right]. \qquad (2.12)$$

Here we have presented the delta function as an integral over an extra field $\mathbf{p}$ and explicitly made Gaussian integration over the force. One can thus see that turbulence is equivalent to the field theory of two interacting fields ($\mathbf{v}$ and $\mathbf{p}$) with large Re corresponding to a strong coupling limit (for incompressible turbulence the pressure is recovered from $\mathrm{div}(\mathbf{v}\cdot\nabla)\mathbf{v} = \Delta P$). For fans of field theory, add that the convective derivative $\mathrm{d}/\mathrm{d}t = \partial/\partial t + (\mathbf{v}\cdot\nabla)$ can be identified as a covariant derivative in the framework of a gauge theory; here the velocity of the reference frame fixes the gauge.

2.2.3 Turbulent flows

The next step is from idealized isotropic homogeneous turbulence to turbulent flows which are necessarily anisotropic and inhomogeneous. Let us apply the new knowledge of turbulence as a multiscale flow to the large-Reynolds flows down an inclined plane and past the body.

River and pipe. Now that we know that turbulence makes the drag at large Re much larger than the viscous drag for a laminar flow, we can understand why the behavior of real rivers is so distinct from a laminar solution from Section 1.4.3. Denote the mean velocity as U. At small Re, the momentum injected by gravity into the fluid is carried to the bottom by the viscous momentum flux, that is the gravity force (per unit mass) $g\alpha$ is balanced by the viscous drag $\nu U/h^2$, or equivalently the input power $g\alpha U$ is equal to the rate of the viscous dissipation vU^2/h^2. At large Re, we expect turbulence with the energy dissipation rate and drag force independent of the viscosity. Then the power $g\alpha U$ must be equal to the dissipation rate which can be estimated as U^3/h, and the drag U^2/h balances $g\alpha$ so that

$$U \simeq \sqrt{\alpha g h}. \tag{2.13}$$

Indeed, as long as viscosity does not enter into things, this is the only combination with the velocity dimensionality that one can get from h and the effective gravity αg. For slow plain rivers (inclination angle $\alpha \simeq 10^{-4}$ and depth $h \simeq 10\,\mathrm{m}$), the new estimate (2.13) gives a reasonable $U \simeq 10\,\mathrm{cm\,s}^{-1}$. Another way to describe the drag is to say that molecular viscosity v is replaced by turbulent viscosity $v_T \simeq Uh \simeq vRe$ and the drag is still given by the viscous formula vU/h^2 but with $v \to v_T$. Intuitively, one imagines turbulent eddies transferring momentum between fluid layers.

Similar arguments can be applied to flows in pipes and channels under the action of the pressure gradient, replacing in (2.13) αg by $\nabla P/\rho$. One is then tempted to conclude that the dimensionless friction factor, defined as $\nabla Ph/\rho U^2$ or $\alpha g h/U^2$, decreases with Re as $1/Re$ at small Re and saturates to a constant at large Re, just like the drag coefficient shown in Figure 1.18. Here h is the channel half-width or pipe radius.

A closer look reveals, however, a flaw in the argument leading to (2.13) – it treats the mean flow and turbulence as homogeneous, while they must be z-dependent to carry the momentum injected by gravity or pressure gradient toward the bottom or walls to be absorbed there. Let us write the momentum conservation without assuming the flow unidirectional. Denote the velocity x-component as $U(z) + u(x,y,z,t)$ and z-component as $v(x,y,z,t)$, where apparently u,v describe turbulent fluctuations. Then the continuity equation for the x-component of the mean momentum states that the divergence of the momentum flux τ is equal to the force:

$$\frac{d}{dz}\left(v\frac{dU}{dz} + \langle uv\rangle\right) \equiv \frac{d\tau(z)}{dz} = -\alpha g h. \tag{2.14}$$

Integrating we get $\tau(z) = \tau(0) - \alpha g z$. The flux is zero on the river surface or at the center of a pipe, which gives $\tau(0) = \alpha g h$. Let us now consider the flow at $z \ll h$ where the momentum flux can be considered independent of z, and denote $v_*^2 \equiv \tau(0) = \alpha g h$. In this region the mean velocity is independent of h and must depend only on v, z, v_*. By dimensional reasoning the dependence must have a form $U = v_* f(zv_*/v)$. The dimensionless parameter zv_*/v is the Reynolds number with the scale set by the distance to the solid boundary. Near the boundary, viscosity absorbs the flux: $vdU/dz = \alpha g$ and $U(z) = \alpha g h z/v$.

The width of that viscous boundary layer can be estimated requiring the Reynolds number to be of order unity: $l = v/v_*$. Outside of this layer, for $z \gg l$, one may expect viscosity to be unimportant and the flux carried by turbulence. As we cannot yet develop a consistent theory of such inhomogeneous turbulence, let us use plausible arguments. Since there is no momentum flux in a uniform flow, then it is natural to relate the mean flux to the mean velocity gradient, dU/dz, which must be determined solely by v_* and z at $l \ll z \ll h$. The only dimensionally possible relation is $dU/dz \simeq v_*/z$, which gives a logarithmic velocity profile for turbulent boundary layer (Karman 1930, Prandtl 1932):

$$U(z) \simeq v_* \log(z/l) = \sqrt{\alpha g h} \log[z(\alpha g h)^{1/2}/v]. \tag{2.15}$$

We used l to make the argument of the logarithm dimensionless since for $z \simeq l$ one must have $U(l) \simeq v_*$. One can further illuminate the hypothesis underlying the log law (2.15) using so-called overlap argument. The dimensionless quantity $U(z)/v_*$ must be a function of two dimensionless arguments, $\ell = z/h$ and $Re = v_* h/v$. Near the wall we expect h to disappear: $U(z)/v_* \rightarrow f(\ell Re)$. Near the center, we expect v to disappear from the law of the velocity change: $U(h) - U(z) = v_* f_1(\ell)$. Denote $U(h)/v_* = f_2(Re)$. We now make an assumption that the two asymptotic regions overlap. In this overlap region we have $f(\ell Re) = f_2(Re) - f_1(\ell)$, which requires all the functions to be logarithmic. Logarithmic turbulent profile is more flat than parabolic laminar profile, which is natural since turbulence better mixes momentum. The overlap argument and claim that the momentum flux completely determines the mean flow in a turbulent boundary layer are curiously similar to assuming inertial interval with the energy flux determining everything in the cascade picture of homogenous turbulence. We now know that Kolmogorov–Obukhov theory correctly describes only the third moment of the velocity statistics, while other moments depend on the large scale. It is not yet clear whether the Prandtl–Karman theory must be modified in a similar way. Experiments support logarithmic mean flow profile but show that turbulence statistics depends on h even at $z \ll h$.

We see that (2.15) corrects (2.13) by a viscosity-dependent logarithmic factor. That makes velocity everywhere, even outside of the viscous layer, dependent on viscosity. While this dependence is very slow and for most cases negligible, conceptually it has dramatic consequences. It tells us that when viscosity goes to zero, the width l of the viscous layer shrinks to zero but $U(l) \simeq v_*$ i.e. stays finite. That means that we have an effective slip on the solid boundary. At any finite z, the velocity $U(z)$ goes to infinity, so that the friction factor goes to zero at $v \to 0$ as $\log^{-2}(hv_*/v)$. All this is because we consider the boundary straight and smooth, which explains the dramatic difference from the flow past a body, where curved surface provides for separation of the boundary layer and resulting wake provides for a finite drag coefficient. It is then reasonable to assume that if the logarithmic decrease of the friction factor with the Reynolds number takes place, it stops when l is getting comparable to the size r of the boundary inhomogeneities (which again save the day, as in Section 1.4.4, but in a different way). When $v < rv_*$ one cannot assume the mean flow to be parallel to the solid boundary. Every inhomogeneity then provides its own wake with a finite drag so that $U(r) \simeq v_*$ and the friction factor saturates at $\log^{-2}(h/r)$.

Wake. Let us now describe the entire wake behind a body at $Re = uL/v \gg 1$. Since Re is large, Kelvin's theorem holds outside the boundary layer – every streamline keeps its vorticity. Streamlines are thus divided into those of zero and non-zero vorticity. A separated region of rotational flow (wake) can exist only if streamlines don't go out of it. Yet zero-vorticity streamlines may come in so that the wake grows as one goes away from the body. Velocity is lower in the wake than outside, and instability of the Kelvin–Helmholtz type makes the boundary of the wake wavy. Oscillations then must also be present in the velocity field in the immediate outside vicinity of the wake. Still, only large-scale harmonics of turbulence are present in the outside region, because the flow is potential ($\Delta \phi = 0$) so when it changes periodically along the wake it decays exponentially with distance from the wake boundary. The smaller the scale, the faster it decays away from the wake. Therefore, all the small-scale motions and all the dissipation are inside the turbulent wake. The boundary of the turbulent wake fluctuates in time. In the snapshot sketch in Figure 2.8, the wake is dark, broken lines with arrows are streamlines; see Figure 1.16 for a real wake photo.

Let us describe the time-averaged position of the wake boundary $Y(x)$. The average angle between the streamlines and the x-direction is $v(x)/u$, where $v(x)$ is the rms turbulent velocity, which can be obtained from the condition that the momentum flux through the wake must be x-independent since it is equal to the drag force $F \simeq \rho uvY^2$, as in (1.67). Then

$$\frac{dY}{dx} = \frac{v(x)}{u} \simeq \frac{F}{\rho u^2 Y^2},$$

so that

$$Y(x) \simeq \left(\frac{Fx}{\rho u^2}\right)^{1/3}, \qquad v(x) \simeq \left(\frac{Fu}{\rho x^2}\right)^{1/3}.$$

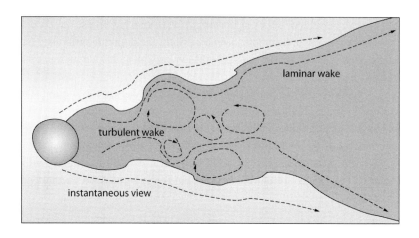

Figure 2.8 Sketch of the wake behind a body.

One can substitute $F \simeq \rho u^2 L^2$ and get

$$Y(x) \simeq L^{2/3} x^{1/3}, \qquad v(x) \simeq u(L/x)^{2/3}.$$

Note that Y is independent of u for a turbulent wake. The current Reynolds number, $Re(x) = v(x)Y(x)/v \simeq (L/x)^{1/3} uL/v = (L/x)^{1/3} Re$, decreases with x and a turbulent wake turns into a laminar one at $x > LRe^3 = L(uL/v)^3$ – the transition distance apparently depends on u.

Inside the laminar wake, under the assumption $v \ll u$, we can neglect $\rho^{-1}\partial p/\partial x \simeq v^2/x$ in the steady Navier–Stokes equation, which then turns into the (parabolic) diffusion equation with x playing the role of time:

$$u\frac{\partial v_x}{\partial x} = v\left(\frac{\partial^2}{\partial z^2} + \frac{\partial^2}{\partial y^2}\right) v_x. \qquad (2.16)$$

At $x \gg v/u$, the solution of this equation acquires the universal form

$$v_x(x,y,z) = -\frac{F_x}{4\pi\eta x} \exp\left[-\frac{u(z^2+y^2)}{4vx}\right],$$

where we have used (1.67) in deriving the coefficient. A prudent thing to ask now is why we accounted for the viscosity in (2.16) but not in the stress tensor (1.65). The answer is that $\sigma_{xx} \propto \partial v_x/\partial x \propto 1/x^2$ decays fast with the distance while $\int dy \sigma_{yx} = \int dy \partial v_x/\partial y$ vanishes identically.

We see that the laminar wake width is $Y \simeq \sqrt{vx/u}$, that is the wake is parabolic. The Reynolds number further decreases in the wake according to the law $v_x Y/v \propto x^{-1/2}$. Recall that in the Stokes flow $v \propto 1/r$ only for $r < v/u$, while in the laminar wake $v_x \propto 1/x$ ad infinitum. Comparing laminar and

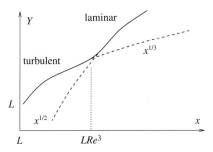

Figure 2.9 Wake width Y versus distance from the body x.

turbulent estimates, we see that for $x \ll LRe^3$, the turbulent estimate gives a larger width: $Y \simeq L^{2/3}x^{1/3} \gg (vx/u)^{1/2}$. On the other hand, in a turbulent wake the width grows and the velocity perturbation decreases with distance slower than in a laminar wake (Figure 2.9).

Wake consideration can be readily generalized for an arbitrary space dimensionality d, where $F \simeq \rho uvY^{d-1}$, so that $v \propto Y^{1-d}$ and $Re \propto vY \propto Y^{2-d}$. We see that wake expansion does not cause the decrease of the Reynolds number in two dimensions. For a body extended across the flow (like a long wing), one can consider the wake two-dimensional and the Reynolds number approximately constant until the distance comparable to the body span.

2.2.4 Mixing

The diffusivity of gases in gases is $\kappa \simeq 10^{-1} cm^2 s^{-1}$ and liquids in liquids $\kappa \simeq 10^{-5} cm^2 s^{-1}$, so it would take many hours for a smell to diffuse across the dinner table and milk across the coffee cup. It is the motion of fluids that usually provides mixing. Let us denote θ the density of an additive. It satisfies the continuity equation which for an incompressible flow turns into the advection-diffusion equation:

$$\frac{\partial \theta}{\partial t} = -\text{div}\,\mathbf{J} = -\text{div}\,(\mathbf{v}\theta - \kappa\nabla\theta) = -(\mathbf{v}\cdot\nabla)\theta + \text{div}\,\kappa\nabla\theta. \qquad (2.17)$$

To show that incompressible flows can only increase the flux of θ, consider two surfaces where we keep different values θ_1, θ_2 and the normal velocity zero (Zeldovich 1937). We assume that all the flux J generated by one surface is absorbed by another, multiply (2.17) by θ and integrate over the space outside the surfaces:

$$\frac{\partial}{\partial t} \int \frac{\theta^2}{2}\, dV = J(\theta_1 - \theta_2) - \int \kappa |\nabla \theta|^2\, dV . \qquad (2.18)$$

Here the identity $\theta(\mathbf{v} \cdot \nabla)\theta = (\mathbf{v} \cdot \nabla)\theta^2/2 = \text{div}\,(\mathbf{v}\theta^2/2)$ made the velocity contribution to disappear. We now consider a steady state or average over time so that the term with time derivative disappears. Then, for a given θ_1, θ_2, the flux is proportional to the integral of dissipation:

$$J(\theta_1 - \theta_2) = \int \kappa |\nabla \theta|^2\, dV . \qquad (2.19)$$

The minimum of the integral with respect to the variations of θ that vanish on the boundaries is achieved by the solution of the equation $\text{div}\,\kappa\nabla\theta = 0$, which corresponds to time-independent or time-averaged solution of (2.17) with zero velocity.[8]

The simplest example of the enhancement of the molecular diffusion due to its interplay with a flow is given by spreading of an additive as it moves along narrow pipes and channels containing uni-directional laminar flow. Consider timescales much exceeding the time of diffusion across the pipe, a^2/κ, where a is the pipe radius or the channel half-width. The velocity and the concentration are non-uniform *across* the pipe, which brings extra diffusion *along* the pipe. The effective longitudinal diffusivity can be estimated as the product of the velocity difference between the center and the wall, which we denote U, and the size of the region of inhomogeneity of θ. As the spot of θ spreads along the pipe, molecular diffusion makes it uniform except two intervals at each end, which appeared most recently, during the time not exceeding a^2/κ. The size of these regions of inhomogeneity can thus be estimated as Ua^2/κ. Multiplying this length by U we obtain an addition to the diffusivity of order U^2a^2/κ (Taylor 1953). That estimate can be validated by deriving from (2.17) the equation that describes the diffusion along the pipe, see Exercise 2.3. The dimensionless parameter $Pe = Ua/\kappa$ is called the Peclet number, it measures relative importance of flow and molecular diffusion. The effective diffusivity $\kappa + kU^2a^2/\kappa$ as a function of κ has a minimum. Here k is a dimensionless factor determined by the geometry of the channel, for a circular tube $k = 1/192$. You can now estimate how fast a drop of medicine spreads in your bloodstream after intravenous injection: taking $a = 0.2\,\text{cm}$, $U = 0.5\,\text{cm/s}$ and $\kappa \simeq 10^{-5}\,\text{cm}^2/\text{s}$, we obtain $Pe = Ua/\kappa = 10^4$, that is spreading is dominated by Taylor dispersion and the effective diffusivity is $U^2a^2/192\kappa = \kappa Pe^2/192 \simeq 10\,\text{cm}^2/\text{s}$. In one second the drop shifts by $0.25\,\text{cm}$ and spreads approximately by $3\,\text{cm}$.

Another example of diffusivity enhancement at large Pe is seen for diffusion through vortices, for instance, convective cells, which we also characterize by their velocity U and size a. To pass to the next cell, the substance must diffuse across a separatrix. The width ℓ of the boundary layer across the separatrix can be estimated requiring the diffusion time ℓ^2/κ to be of order of the turnover time a/U, which gives $\ell \simeq (\kappa a/U)^{1/2}$. Since only fluid particles from the boundary layers are able to cross to the next cell and travel far (see the Figure), then that width plays the role of the mean free path, and the enhanced diffusivity is $U\ell \simeq \sqrt{\kappa U a} = \kappa Pe^{1/2}$.

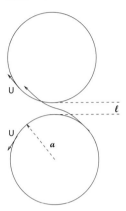

Let us consider now turbulent velocity field $\mathbf{v}(\mathbf{r},t)$ and ask how far a spot of θ may deviate from the mean flow during time t. The coordinate of the spot center satisfies the equation

$$\frac{d\mathbf{q}}{dt} = \mathbf{v}\left[\mathbf{q}(t),t\right] ,$$

whose solution is the integral of the Lagrangian velocity $\mathbf{V}(t) = \mathbf{v}\left[\mathbf{q}(t),t\right]$ over time: $\mathbf{q}(t) = \mathbf{q}(0) + \int_0^t \mathbf{V}(t')\,dt'$. If the correlation time τ_c of $V(t)$ is finite, then at $t \gg \tau_c$ the variance grows by the diffusion law $\langle q_i(t) g_j(t) \rangle = 2D_{ij}t$ where the so-called eddy diffusivity is as follows

$$D_{ij} = \frac{1}{2} \int_0^\infty \langle V_i(0)V_j(s) + V_j(0)V_i(s) \rangle \, ds .$$

Apart from random wandering, the spot also spreads. The spreading of the spot depends on how fast fluid particles separate, which is determined by the dependence of their relative velocity $\delta\mathbf{v}$ on the separation between them, $\mathbf{R} = \mathbf{q}_1 - \mathbf{q}_2$. Two qualitatively different classes of flows must be distinguished: spatially smooth with $\delta v(R) \propto R$ and nonsmooth with $\delta v(R) \propto R^{1-\alpha}$ and $0 < \alpha < 1$. As we learned in Section 2.2.2, turbulent flows are spatially smooth at the scales smaller than the viscous scale l (viscous interval) and non-smooth at larger scales (inertial interval), where $\delta v(R) \propto R^{1/3}$. On the viscous scale, $\delta v(l)l \simeq \nu$, so that the Peclet number is ν/κ. That ratio is called Schmidt number or Prandtl number when θ is temperature. Even though the momentum and substance diffusivities are caused by the same molecular motion, their ratio widely varies depending on the type of material (see also Exercise 2.2). The Schmidt number ν/κ is very high for viscous liquids and also for colloids and aerosols since the diffusivity of, say, micron-size cream globules in milk and smoke in the air is six to seven orders of magnitude less than the viscosity of the ambient fluid. In those cases, flows dominate spreading.

At the scales less than l, the velocity difference can be linearized in distance i.e. completely characterized by the matrix of the velocity derivatives $\nabla_i v_k$, as in Section 1.2.2 and Exercise 2.1. When the antisymmetric (vorticity) part of this matrix dominates, fluid elements rotate. In strain-dominated regions fluid elements are deformed: stretched and contracted. The net result of a long sequence of random events of stretching, contraction and rotation is turning any ball into an elongated ellipsoid. The physical reason for it is that substantial deformation appears sooner or later. To reverse it, one needs to contract the long axis of the ellipsoid, that is the direction of contraction must be inside the narrow angle defined by the eccentricity. It is less likely to meet a deformation inside rather than outside the angle, so that randomly oriented deformations on average continue to increase the eccentricity. As a result, fluid particles which started close to each other can be found far away, which leads to mixing. Drop ink into a glass of water, gently stir (not shake) and enjoy the visualization of exponential stretching and mixing.

Formally, the evolution of the inter-particle distance in a smooth flow is linear: $\mathbf{R}(t) = W\mathbf{R}(0)$. The positive symmetric matrix $t^{-1} \ln W^T W$ stabilizes as $t \to \infty$: eigenvectors tend to d fixed orthonormal vectors $\mathbf{f}_i$, and the limiting eigenvalues, $\lambda_i = \lim_{t \to \infty} t^{-1} \ln |W\mathbf{f}_i|$, do not depend on the starting point if the flow is ergodic. The mean stretching/contraction rates λ_i are called Lyapunov exponents. In an incompressible flow $\sum_i \lambda_i = 0$, so in a generic case there exists at least one positive exponent which gives stretching.

The directions that contract are eventually stopped by molecular diffusion. Yet the exponentially growing directions continue to expand, so that the volume grows exponentially and the value of θ inside the spot decays exponentially in time. For an arbitrary large-scale initial distribution of θ, the concentration variance decays exponentially in a spatially smooth flow since this is how fast velocity inhomogeneity contracts θ "feeding" molecular diffusion which eventually decreases the variance. At $Pe \gg 1$, even though it is diffusion that diminishes θ, the rate of decay is usually of order of the typical velocity gradient that is independent of κ (another example of dissipative anomaly). Recall that vorticity and magnetic field satisfy the same equation (1.22) as the distance between two close fluid particles, so they can also be exponentially stretched; in particular, this is the mechanism of magnetic dynamo in the Earth core and interstellar gas.

At the scales larger than l, in the inertial interval of turbulence cascade, the velocity difference (2.8) scales as $\delta v(r) \propto r^{1/3}$ and the particles separate according to the Richardson's law $R(t) \propto t^{3/2}$, as we learnt in Section 2.2.2. This is faster than both diffusion ($R \propto t^{1/2}$) and ballistics ($R \propto t$). The reason for it is a multiscale nature of turbulence, that is the presence of vortices of different sizes. We expect only vortices with $r < R$ to participate in separation, since larger vortices just sweep both particles together. As the particles separate, vortices of larger r and larger $\delta v(r)$ participate and accelerate separation. The volume of any spot also grows so that the scalar variance decays by a power law: $\langle \theta^2 \rangle \propto t^{-3d/2}$, where d - space dimensionality.

Since the velocity difference (2.8) increases with the distance slower than linearly, then the velocity in turbulence is non-Lipschitz on average, see Section 1.1, so that fluid trajectories are not well-defined in the inviscid limit.[9]

Considering two fluid particles whose relative velocity has a negative projection on the line connecting them, and solving the equation $dR/dt = -CR^{1/3}$ we obtain $R^{2/3}(t) = R^{2/3}(0) - 3Ct/2$, which suggests that the trajectories may intersect in a finite time. Does that mean that if we create a smooth incompressible initial flow with large Re, appearance of turbulence will lead to velocity nonsmoothness in a finite time? Can we claim our Clay prize now? Not just yet. In such a local flow configuration, the current Reynolds number $R\delta v(R)/\nu \propto R^{4/3}$ decreases with the distance so that the viscosity will stop this at the scale l where cascade stops. But Richardson's law and (2.8) are expected to describe some mean properties of turbulent flows. Maybe finite-time singularities appear in rare fluctuations that do not contribute the cascade? Can we imagine such local flow configuration that compresses energy into a region of diminishing size $\ell(t)$ keeping the Reynolds number $v\ell/\nu$ large? Requiring the energy (rather than the energy flux) constant, we obtain $v^2 \ell^d = E =$const. That suggests $v = E^{1/2}\ell^{-d/2}$, so that $d\ell/dt = -v$ gives another law of turning the scale into zero: $\ell^{1+d/2}(t) = \ell^{1+d/2}(0) - (1+d/2)E^{1/2}t$. Let us check if the viscosity can stop such a contraction: $Re(t) = v(t)\ell(t)/\nu \propto \ell^{1-d/2}$. The case of $d = 2$ is marginal but it has extra conservation laws which make finite-time singularity impossible. But for $d = 3$ the Reynolds number is expected to grow upon an energy-conserving contraction so that the viscosity is getting irrelevant and cannot stop it. We thus see that the energy conservation does not forbid a finite-time singularity in three dimensions, but so far nobody was able to find such a flow or show that it is impossible.

While incompressible flows generally mix, compressible flows can segregate. In other words, θ differences are decreased by the former while they can be increased by the latter. We shall consider dynamics of compressible flows in the next section, here we make few remarks on kinematics of the fluid particles and substances they carry. The continuity equation written in a Lagrangian form in the reference frame moving with the flow is as follows

$$\frac{d\theta}{dt} = -\theta\,\mathrm{div}\,\mathbf{v}\,, \quad \ln\left[\theta(t)/\theta(0)\right] = \int_0^t \mathrm{div}\,\mathbf{v}\left[\mathbf{q}(t'),t'\right]dt' = C(t)\,. \quad (2.20)$$

Volume conservation means that the compression factor averaged over the whole flow is zero: $\langle C \rangle = 0$. However, the *concavity* of the exponential function means that the average exponent of it is generally larger than unity: $\langle [\theta(t)/\theta(0)] \rangle = \langle e^C \rangle \geq 1$. This is because the parts of the flow with positive $C(t)$ give more contribution into the exponent than the parts with negative $C(t)$. Moreover, for a fluctuating flow, $\langle [\theta(t)/\theta(0)] \rangle$ generally grows with time. Indeed, if the Lagrangian quantity $\mathrm{div}\,\mathbf{v}\left[\mathbf{q}(t),t\right]$ is a random function with zero mean and a finite correlation time, then at longer times the logarithm of the density must have a Gaussian statistics with zero mean and the variance linearly growing with time. That means that fluctuations grow, so that the distribution is getting more and more nonuniform. The Eulerian picture is that the more and more of the space is emptied, while the Lagrangian picture is that more and more fluid elements are getting compressed into a fractal smeared by molecular diffusion.[10] This is to be contrasted with the mixing and decay of inhomogeneities in an incompressible flow with molecular diffusion.

An interesting and important example of a compressible flow is that of a cloud of particles suspended in a fluid, for instance, rain and cloud droplets in the air. When there are many such particles we may consider the set of their velocities as a field $\mathbf{u}(\mathbf{r},t)$ and treat the equation (1.61) as a partial differential equation relating this field to the fluid velocity field $\mathbf{v}(\mathbf{r},t)$:

$$\frac{d\mathbf{u}}{dt} = \frac{\partial\mathbf{u}}{\partial t} + (\mathbf{u}\cdot\nabla)\mathbf{u} = \frac{\mathbf{v}-\mathbf{u}}{\tau}\,.$$

Even when $\mathrm{div}\,\mathbf{v} = 0$ one obtains for $\sigma(t) = \mathrm{div}\,\mathbf{u}$

$$\frac{d\sigma}{dt} + \frac{\sigma}{\tau} = -\nabla_i v_k \nabla_k v_i = -\frac{1}{4}\left[(\nabla_i v_k + \nabla_k v_i)^2 - (\nabla_i v_k - \nabla_k v_i)^2\right] = \omega^2 - S^2\,.$$

We see that the expansion rate σ is positive in vorticity-dominated elliptic regions. That means that droplets are thrown out of the air vortices by centrifugal force. The droplets concentrate in strain-dominated hyperbolic regions. Segregation of inertial particles in random flows have many consequences, from planet formation to rain initiation.

The above considerations assumed randomness for Lagrangian quantities, which does not necessarily require flow fluctuating in time, it is enough to have flow pattern complicated enough to provide for the so-called Lagrangian chaos. Indeed, the system of d nonlinear ordinary differential equations, $d\mathbf{q}/dt = \mathbf{v}(\mathbf{q})$, for generic $\mathbf{v}(\mathbf{q})$ has a complicated phase portrait with stochastic attractors exactly like in dynamical chaos mentioned in Section 2.2.1.

The take home lessons of this Section: i) how exponential growth of perturbations in time and exponential separation of trajectories in space make statistics to appear from dynamics, ii) how the effect of symmetry breaking can stay finite when the symmetry breaking factor goes to zero.

2.3 Acoustics

Another set of unsteady phenomena is related to the finiteness of the speed of sound c. We first consider fluid velocity v much less than c and describe linear acoustics (in the first order in v/c). We then account for small nonlinearity and dissipation and introduce the Burgers equation. Finally, we consider phenomena that appear when v exceeds c.

2.3.1 Sound

Small perturbations of density in an ideal fluid propagate as sound waves that are described by the continuity and Euler equations linearized with respect to the perturbations $p' \ll p_0, \rho' \ll \rho_0$:

$$\frac{\partial \rho'}{\partial t} + \rho_0 \mathrm{div}\,\mathbf{v} = 0, \qquad \frac{\partial \mathbf{v}}{\partial t} + \frac{\nabla p'}{\rho_0} = 0. \tag{2.21}$$

To close the system we need to relate the variations of the pressure and density, i.e. specify the equation of state. The derivative of the pressure with respect to the density has the dimensionality of velocity squared so we denote it c^2, then $p' = c^2 \rho'$. As noticed in Section 1.2.4, small oscillations are potential so we introduce $\mathbf{v} = \nabla \phi$ and get from (2.21)

$$\phi_{tt} - c^2 \Delta \phi = 0. \tag{2.22}$$

We see that indeed c is the velocity of sound. What is left to establish is what kind of derivative $\partial p/\partial \rho$ one uses, isothermal or adiabatic, which differ substantially. For a gas, the isothermal derivative gives $c^2 = P/\rho$ while the adiabatic law $P \propto \rho^\gamma$ gives:

$$c^2 = \left(\frac{\partial p}{\partial \rho}\right)_s = \frac{\gamma p}{\rho}. \tag{2.23}$$

One uses an adiabatic equation of state when one can neglect the heat exchange between compressed (warmer) and expanded (colder) regions. This means that the thermal diffusivity (estimated as thermal velocity times the mean free path) must be less than the sound velocity times the wavelength. Since the sound velocity is of the order of the thermal velocity, it requires the wavelength to be longer than the mean free path, which is always so. Therefore, sound must be always treated as adiabatic. Newton already knew that $c^2 = \partial p / \partial \rho$. Experimental data from Boyle showed $p \propto \rho$ (i.e. they were isothermal), which suggested for air $c^2 = p / \rho \simeq 290$ m s^{-1}, well off the observed value of 340 m s^{-1} at $20\,°C$. It was only a hundred years later that Laplace got the true (adiabatic) value with $\gamma = 7/5$.

All velocity components, pressure and density perturbations also satisfy the *wave equation* (2.22). A particular solution of this equation is a monochromatic plane wave, $\phi(\mathbf{r},t) = \cos(i\mathbf{k}\mathbf{r} - i\omega t)$. The relation between the frequency ω and the wavevector $\mathbf{k}$ is called the dispersion relation; for acoustic waves it is linear: $\omega = ck$. In one dimension, the general solution of the wave equation is particularly simple:

$$\phi(x,t) = f_1(x - ct) + f_2(x + ct), \tag{2.24}$$

where f_1, f_2 are given by two initial conditions, for instance, $\phi(x,0)$ and $\phi_t(x,0)$. Note that only $v_x = \partial \phi / \partial x$ is nonzero so that sound waves in fluids are longitudinal. Any localized 1D initial perturbation (of density, pressure or velocity along x) thus breaks into two plane-wave packets moving in opposite directions without changing their shape. In every such packet, $\partial / \partial t = \pm c \partial / \partial x$ so that the second equation (2.21) gives

$$v = p'/\rho c = c\rho'/\rho. \tag{2.25}$$

The wave amplitude is small when $\rho' \ll \rho$, which requires $v \ll c$. The (fast) pressure variation in a sound wave, $p' \simeq \rho v c$, is much larger than the (slow) variation $\rho v^2 / 2$ one estimates from the Bernoulli theorem.

Considering spherically symmetric case in d dimensions, one finds that the equation

$$\phi_{tt} = \frac{c^2}{r^{d-1}} \frac{\partial}{\partial r} \left(r^{d-1} \frac{\partial \phi}{\partial r} \right) \tag{2.26}$$

turns into $h_{tt} = c^2 \partial^2 h / \partial r^2$ with the substitution $\phi = h/r^a$ only for $d = 1, a = 0$ and $d = 3, a = 1$. We can thus find the general solution of (2.26) in three dimensions as well:

$$\phi(r,t) = r^{-1}[f_1(r - ct) + f_2(r + ct)]. \tag{2.27}$$

For the case of axial symmetry see the Problem 2.9.

The energy density of sound waves E_w can be obtained by expanding $\rho E + \rho v^2/2$ up to the second-order terms in perturbations. The zero-order term $\rho_0 E_0$ is constant and thus unrelated to waves. The first-order term $\rho' \partial(\rho E)/\partial \rho = w_0 \rho'$ is related to the mass redistribution. While this term is generally nonzero, it disappears after integration over the whole volume, so we omit it from the wave energy. We are left with the quadratic terms:

$$E_w = \frac{\rho_0 v^2}{2} + \frac{\rho'^2}{2}\frac{\partial^2(\rho E)}{\partial \rho^2} = \frac{\rho_0 v^2}{2} + \frac{\rho'^2}{2}\left(\frac{\partial w_0}{\partial \rho}\right)_s = \frac{\rho_0 v^2}{2} + \frac{\rho'^2 c^2}{2\rho_0}.$$

The energy flux with the same accuracy is

$$\mathbf{q} = \rho \mathbf{v}(w + v^2/2) \approx \rho \mathbf{v} w \approx w' \rho_0 \mathbf{v} + w_0 \rho' \mathbf{v}.$$

We must omit the energy flux caused by the mass change in a given volume, $w_0 \rho' \mathbf{v}$, since it corresponds to the term $w_0 \rho'$ omitted in the energy. The enthalpy variation is $w' = p'(\partial w/\partial p)_s = p'/\rho \approx p'/\rho_0$ and we obtain

$$\mathbf{q} = p' \mathbf{v}.$$

The energy and the flux are related by $\partial E_w/\partial t + \mathrm{div}\, p'\mathbf{v} = 0$. For a plane wave, we obtain from (2.25) $E_w = \rho_0 v^2$ and $q = cE_w$. The energy flux is also called the acoustic intensity. To better hear the distant murmur of the brook and be less frightened by the sudden lion's roar, our ear amplifies weak sounds and damp strong ones. It does that by sensing loudness as the logarithm of the intensity for a given frequency. This is why the acoustic intensity is traditionally measured not in watts per square metre but in units of the intensity logarithm, called decibels:

$$q(\mathrm{dB}) = 120 + 10\log_{10} q \ (\mathrm{W/m^2}).$$

Normal conversation in most countries is about 50–60 dB, rock concert is over 100 dB.

The momentum density is

$$\mathbf{j} = \rho \mathbf{v} = \rho_0 \mathbf{v} + \rho' \mathbf{v} = \rho_0 \mathbf{v} + \mathbf{q}/c^2.$$

Propagating acoustic perturbation of a finite extent in a volume not restricted by walls has a nonzero total momentum [11]

$$\int \mathbf{j}\, dV = \rho_0 \int \nabla \phi\, dV + \int \mathbf{q}\, dV/c^2 = \int \mathbf{q}\, dV/c^2 . \qquad (2.28)$$

Comment briefly on the momentum of a phonon, which is defined as a sinusoidal perturbation of atom displacements; a monochromatic wave in these (Lagrangian) coordinates has zero momentum. [12] But our equations (2.22,2.26) and solutions (2.24,2.27) are written in Eulerian coordinates. A perturbation, which is sinusoidal in Eulerian coordinates, has a nonzero momentum at second order (where Eulerian and Lagrangian coordinates differ). Indeed, let us consider the Eulerian velocity field as a monochromatic wave with a given frequency and a wavenumber: $v(x,t) = u\sin(kx - \omega t)$. The Lagrangian coordinate $X(t)$ of a fluid particle satisfies the following equation:

$$\dot{X} = v(X,t) = u\sin(kX - \omega t). \tag{2.29}$$

This is a nonlinear equation, which can be solved by iteration, $X(t) = X_0 + X_1(t) + X_2(t) + \ldots$ assuming $v \ll \omega/k$. We shall see below in Section 3.1.4 that ω/k is the wave phase velocity. When it is much larger than the fluid velocity, the fluid particle displacement during the wave period is much smaller than the wavelength. Such an iterative solution gives oscillations at first order and a mean drift at second order:

$$X_1(t) = \frac{u}{\omega}\cos(kX_0 - \omega t),$$
$$X_2(t) = \frac{ku^2 t}{2\omega} - \frac{ku^2}{4\omega^2}\sin 2(kX_0 - \omega t). \tag{2.30}$$

We see that at first order in wave amplitude the perturbation propagates, while at second order the fluid itself flows in the direction of wave propagation with the speed $ku^2/2\omega = u^2/2c$. Fluid particles move with the wave a bit longer than against it.

However, nonzero momentum does not necessarily mean net flow in one dimension. Fluid particles exchange momentum, which thus can be transported without mass transfer. For example, one can generate the wave train by moving piston in a tube back and forth, without producing any net flow. In this case, every fluid particle oscillates in the wave and returns to its original position after the wave train has passed by it. Therefore, the time average of the mass flux must be zero at any point: $\overline{\rho v} = \rho_0 \bar{v} + \overline{\rho' v} = 0$. That requires mean counterflow, $\bar{v} = -\overline{\rho' v}/\rho_0 = -\overline{v^2}/c = -u^2/2c$, which exactly cancels the Lagrangian drift (2.30), derived under the assumption of no mean Eulerian velocity. The lesson is that the drift is a quadratic quantity and must be determined by using the Eulerian velocity valid up to quadratic terms as well. While the momentum $\overline{\rho v}$ averaged over time at a given point is zero, the momentum (2.28) averaged over space at a given time is nonzero.

2.3.2 Riemann wave

As we have seen, an infinitesimally small one-dimensional acoustic perturbation splits into two simple waves, which then propagate without changing their forms. Let us show that such purely adiabatic waves of a permanent shape are impossible for finite amplitudes (Earnshaw paradox): in the reference frame moving with speed c one would have a steady motion with the continuity equation $\rho v = \text{const.} = C$ and the Euler equation $v\,dv = -dp/\rho$, giving $dp/d\rho = (C/\rho)^2$ and $p = p_0 - C^2/\rho$, which contradicts the second law of thermodynamics (such gas would have negative pressure for sufficiently large momentum C or sufficiently small density). It is thus clear that a simple plane wave must change under the action of a small factor of nonlinearity.

Consider now a one-dimensional motion of a compressible fluid having the usual adiabatic equation of state $p = p_0(\rho/\rho_0)^\gamma$ with $\gamma > 1$. Let us look for a simple wave where one can express any two of v, p, ρ via the remaining one. This is a generalization for a nonlinear case of what we did for a linear wave (only now it must be nonstationary). Suppose that we assume everything to be determined by v, that is $p(v)$ and $\rho(v)$. The Euler and continuity equations take the form:

$$\frac{dv}{dt} = -\frac{1}{\rho}c^2(v)\frac{d\rho}{dv}\frac{\partial v}{\partial x}, \qquad \frac{d\rho}{dv}\frac{dv}{dt} = -\rho\frac{\partial v}{\partial x}. \qquad (2.31)$$

Here $c^2(v) \equiv dp/d\rho$. Excluding $d\rho/dv = \pm\rho/c$, one gets

$$\frac{dv}{dt} = \frac{\partial v}{\partial t} + v\frac{\partial v}{\partial x} = \pm c(v)\frac{\partial v}{\partial x}. \qquad (2.32)$$

Two signs correspond to waves propagating in opposite directions. In a linear approximation we had $u_t + cu_x = 0$, where $c = \sqrt{\gamma p_0/\rho_0}$. Now we find

$$
\begin{aligned}
c(v) &= \sqrt{\frac{\gamma p}{\rho}} = \sqrt{\gamma\frac{p_0 + \delta p}{\rho_0 + \delta\rho}} \\
&= c\left(1 + \frac{\delta p}{2p_0} - \frac{\delta\rho}{2\rho_0}\right) = c + v\frac{\gamma - 1}{2},
\end{aligned} \qquad (2.33)
$$

since $\delta\rho/\rho_0 = v/c$. The local sound velocity increases with amplitude since $\gamma > 1$, that is the positive effect of the pressure increase overcomes the negative effect of the density increase.

Taking a plus sign in (2.32) we get the equation for the simple wave propagating rightwards [13]

$$\frac{\partial v}{\partial t} + \left(c + v\frac{\gamma + 1}{2}\right)\frac{\partial v}{\partial x} = 0. \qquad (2.34)$$

This equation describes the simple fact that the higher the amplitude of the perturbation the faster it propagates, both because of higher velocity and of higher pressure gradient. (J. S. Russel remarked in 1885 that "the sound of a

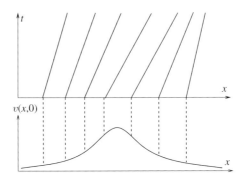

Figure 2.10 Characteristics (upper panel) and the initial velocity distribution (lower panel).

cannon travels faster than the command to fire it"). Since different parts of the wave profile move with different speeds, then the profile changes. In particular, faster fluid particle will catch up with slower moving particles. Indeed, if we have the initial distribution $v(x,0) = f(x)$, the solution of (2.34) is given by an implicit relation

$$v(x,t) = f\left[x - \left(c + v\frac{\gamma+1}{2}\right)t\right],\tag{2.35}$$

which can be useful for particular f but is not much help in a general case. An explicit solution can be written in terms of characteristics (the lines in the $x-t$ plane that correspond to constant v, Figure 2.10):

$$\left(\frac{\partial x}{\partial t}\right)_v = c + v\frac{\gamma+1}{2} \quad \Rightarrow x = x_0 + ct + \frac{\gamma+1}{2}v(x_0)t,\tag{2.36}$$

where $x_0 = f^{-1}(v)$. The solution (2.36) is called a simple or Riemann wave (Riemann 1858).

In the variables $\xi = x - ct$ and $u = v(\gamma+1)/2$, the equation takes the form

$$\frac{\partial u}{\partial t} + u\frac{\partial u}{\partial \xi} = \frac{du}{dt} = 0,$$

which describes freely (inertially) moving particles. Indeed, we see that the characteristics are straight lines with the slopes given by the initial distribution $v(x,0)$, that is every fluid particle propagates with a constant velocity. It is seen that the parts where $\partial v(x,0)/\partial x$ was initially positive will decrease their slope while the negative slopes in $\partial v(x,0)/\partial x$ become steeper (Figure 2.11).

The characteristics are actually Lagrangian coordinates: $x(x_0,t)$. The characteristics cross in the $x-t$ plane (and particles hit each other) when $(\partial x/\partial x_0)_t$ turns into zero, that is

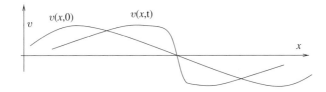

Figure 2.11 Evolution of the velocity distribution toward wave breaking.

$$1 + \frac{\gamma+1}{2}\frac{dv}{dx_0}t = 0,$$

which first happens with particles that correspond to $dv/dx_0 = f'(x_0)$ maximal negative, that is $f''(x_0) = 0$. When characteristics cross, we have different velocities at the same point in space, which corresponds to a discontinuity in the velocity field called shock.

A general remark: notice the qualitative difference between the properties of the solutions of the hyperbolic equation $u_{tt} - c^2 u_{xx} = 0$ and the elliptic equations, say the Laplace equation. As mentioned in Section 1.3.1, elliptic equations have solutions and their derivatives which are regular everywhere inside the domain of existence. On the contrary, hyperbolic equations propagate perturbations along the characteristics and characteristics can cross (when c depends on u or x,t), leading to singularities.

2.3.3 Burgers equation

We thus see that in ideal hydrodynamics nonlinearity makes the propagation velocity depend on the amplitude, which leads to crossing of characteristics and thus to wave breaking: any acoustic perturbation tends to create a singularity (shock) in a finite time. Spatial derivatives of the first order enter the equations of ideal hydrodynamics. Apparently, an account of higher derivatives is necessary near a shock. In this section, we account for the next derivative (the second one), which corresponds to viscosity:

$$\frac{\partial u}{\partial t} + u\frac{\partial u}{\partial \xi} = v\frac{\partial^2 u}{\partial \xi^2}. \tag{2.37}$$

This is the Burgers equation, the first representative of the small family of universal nonlinear equations (we shall meet two other equally famous members, the Korteveg–de-Vries and non-linear Schrödinger equations, in the next chapter, where we account, in particular, for the third derivative in acoustic-like perturbations). Note that v in the Burgers equation is half the kinematic viscosity. Indeed, from the linearized Navier–Stokes and continuity equations one can readily obtain for the frequency of sound: $\omega = \sqrt{c^2 k^2 - \iota v k^3 c} \approx ck - \iota v k^2 / 2$.

The Burgers equation is a minimal model of fluid mechanics: a single scalar field $u(x,t)$ changes in one dimension under the action of inertia and friction. This equation describes wide classes of systems with hydrodynamic-type nonlinearity, $(u\nabla)u$, and viscous dissipation. It can be written in a potential form $u = \nabla\phi$; then $\phi_t = -(\nabla\phi)^2/2 + v\Delta\phi$; in such a form it can be considered in one and two spatial dimensions, where it describes in particular [14] the surface growth under uniform deposition and diffusion: the deposition contribution into the time derivative of the surface height $\phi(r)$ is proportional to the flux per unit area, which is inversely proportional to the area, $[1 + (\nabla\phi)^2]^{-1/2} \approx 1 - (\nabla\phi)^2/2$, as shown in Figure 2.12.

The Burgers equation can be linearized by the Hopf substitution $u = -2v\varphi_\xi/\varphi$:

$$\frac{\partial}{\partial\xi}\frac{\varphi_t - v\varphi_{\xi\xi}}{\varphi} = 0 \Rightarrow \varphi_t - v\varphi_{\xi\xi} = \varphi C'(t),$$

which with the change $\varphi \to \varphi \exp C$ (not changing u) gives the linear diffusion equation:

$$\varphi_t - v\varphi_{\xi\xi} = 0.$$

The initial value problem for the diffusion equation is solved as follows:

$$\varphi(\xi,t) = \frac{1}{\sqrt{4\pi v t}} \int_{-\infty}^{\infty} \varphi(\xi',0) \exp\left[-\frac{(\xi - \xi')^2}{4\pi v t}\right] d\xi' \qquad (2.38)$$

$$= \frac{1}{\sqrt{4\pi v t}} \int_{-\infty}^{\infty} \exp\left[-\frac{(\xi - \xi')^2}{4\pi v t} - \frac{1}{2v}\int_0^{\xi'} u(\xi'',0)\, d\xi''\right] d\xi'.$$

Despite the fact that the Burgers equation describes a dissipative system, it conserves total momentum (as any viscous equation does).[15] If the momentum

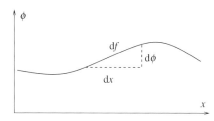

Figure 2.12 If the x-axis is along the direction of the local surface change then the local area element is $df = \sqrt{(dx)^2 + (d\phi)^2} = dx\sqrt{1 + (\nabla\phi)^2}$.

is finite, then any perturbation evolves into a universal form depending only on $M = \int u(x)\,dx$ and not on the form of $u(\xi,0)$. At $t \to \infty$, (2.38) gives $\varphi(\xi,t) \to \pi^{-1/2}F[\xi(4vt)^{-1/2}]$, where

$$F(y) = \int_{-\infty}^{\infty} \exp\left[-\eta^2 - \frac{1}{2v}\int_0^{(y-\eta)\sqrt{4vt}} u(\eta',0)\,d\eta'\right]d\eta$$

$$\approx e^{-M/4v}\int_{-\infty}^{y}e^{-\eta^2}\,d\eta + e^{M/4v}\int_y^{\infty}e^{-\eta^2}\,d\eta. \qquad (2.39)$$

Solutions with positive and negative M are related by the transform $u \to -u$ and $\xi \to -\xi$.

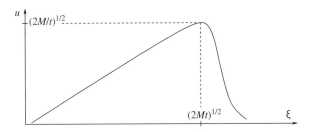

Note that M/v is the Reynolds number and it does not change while the perturbation spreads. This is a consequence of momentum conservation in one dimension. In a free viscous decay of a d-dimensional flow, usually the scale grows as $R(t) \propto t^{1/2}$. To keep the momentum, velocity must decay as $R^{-d} \propto t^{-d/2}$ so that the Reynolds number evolves as $t^{(1-d)/2}$. Compare it with a wake behind the body (see Section 2.2.3), where the Reynolds number does not change at $d=2$. Yet another case is a jet in a fluid (see Section 1.5.2), where constancy of the momentum flux $\int u^2\,df \simeq u^2R^{d-1}$ means that the Reynolds number uR/v does not change along the jet at $d=3$.

When $M/v \gg 1$ the solution looks particularly simple, as it acquires a sawtooth form. Indeed, in the interval $0 < y < M/2v$ (i.e. for $0 < \xi < \sqrt{2Mt}$) the first term in (2.39) is negligible and $F \sim \exp(-y^2)$ so that $u(\xi,t) = \xi/t$. For both $\xi < 0$ and $\xi > \sqrt{2Mt}$ we have $F \sim \text{const.} + \exp(-y^2)$ so that u is exponentially small there.

An example of the solution with an infinite momentum is a steady propagating shock. Let us look for a travelling wave solution $u(\xi - wt)$. In this case the Burgers equation is reduced to an ordinary differential equation which can be immediately integrated to give $-uw + u^2/2 = vu_\xi$ under the assumption that $u \to 0$ at least for one of the infinities. Integrating again:

$$u(\xi,t) = \frac{2w}{1 + C\exp[w(\xi - wt)/v]}. \qquad (2.40)$$

We see that this is a shock having width v/w and propagating with velocity that is half the velocity difference on its sides. A simple explanation is that the shock front is the place where a moving fluid particle hits a standing fluid particle, they stick together and continue with half velocity owing to momentum conservation. The form of the shock front is steady since nonlinearity is balanced by viscosity.

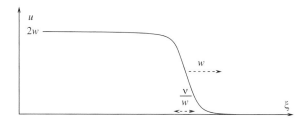

The Burgers equation is Galilean invariant, that is if $u(\xi,t)$ denotes a solution so does $u(\xi - wt) + w$ for an arbitrary w. In particular, one can transform (2.40) into a standing shock, $u(\xi,t) = -w\tanh(w\xi/2v)$.

2.3.4 Acoustic turbulence

The shock wave (2.40) dissipates energy at the rate $v\int u_x^2\,dx$ independent of viscosity, see (2.41). Indeed, when a moving particle sticks to a standing particle, half of their kinetic energy is lost, independently of how sticking actually occurs. In compressible flows, shock creation is a way of dissipating finite energy in the inviscid limit (in incompressible flows, this was achieved by turbulent cascade). The solution (2.40) shows how it works: the velocity derivative goes to infinity as the viscosity goes to zero. In the inviscid limit, the shock is a velocity discontinuity.

Consider now acoustic turbulence produced by a pumping correlated on much larger scales, for example, pumping a pipe from one end by frequencies Ω much less than cw/v, so that the Reynolds number is large. Upon propagation along the pipe, such turbulence evolves into a set of shocks at random positions with the mean distance between shocks $L \simeq c/\Omega$ far exceeding the shock width v/w, which is a dissipative scale. For every shock (2.40),

$$S_3(x) = \frac{1}{L}\int_{-L/2}^{L/2}[u(x+x') - u(x')]^3\,dx' \approx -8w^3x/L,$$

$$\varepsilon = \frac{1}{L}\int_{-L/2}^{L/2}vu_x^2\,dx \approx 2w^3/3L, \qquad (2.41)$$

which gives:

$$S_3 = -12\varepsilon x . \tag{2.42}$$

This formula is a direct analogue of the flux law (2.11). As in Section 2.2.2, it would be wrong to assume $S_n = \langle [u(x) - u(0)]^n \rangle \simeq (\varepsilon x)^{n/3}$, since shocks give a much larger contribution for $n > 1$: $S_n \simeq w^n x/L$, here x/L is the probability of finding a shock in the interval x. In terms of Fourier harmonics, every shock contributes $u_k \propto 1/k$, which indeed gives $S_2(x) = \langle [u(x) - u(0)]^2 \rangle = \int |u_k|^2 (1 - e^{ikx})\, dk \propto \int^{1/x} |u_k|^2\, dk \propto x$.

Generally, $S_n(x) \sim C_n |x|^n + C'_n |x|$, where the first term comes from the smooth parts of the velocity (the right x-interval in Figure 2.13) while the second comes from $O(x)$ probability having a shock in the interval x.

The scaling exponents, $\xi_n = d\ln S_n / d\ln x$, thus behave as follows: $\xi_n = n$ for $0 \leq n \leq 1$ and $\xi_n = 1$ for $n > 1$. Like incompressible (vortex) turbulence in Section 2.2.2, this means that the probability distribution of the velocity difference $P(\delta u, x)$ is not scale-invariant in the inertial interval, that is one cannot find such a that makes the function of the rescaled velocity difference $\delta u/x^a$ scale-independent. The simple bimodal nature of Burgers turbulence (shocks and smooth parts) means that the PDF is actually determined by two (nonuniversal) functions, each depending on a single argument: $P(\delta u, x) = \delta u^{-1} f_1(\delta u/x) + x f_2(\delta u/u_{\mathrm{rms}})$. The breakdown of scale invariance means that the low-order moments decrease faster than the high-order ones as one goes to smaller scales. That means that the level of fluctuations increases with the resolution: the smaller the scale the more probable are large fluctuations. When the scaling exponents ξ_n do not lie on a straight line, this is called an anomalous scaling since it is related again to the symmetry (scale invariance) of the PDF broken by pumping and not restored even when $x/L \to 0$.

As an alternative to (2.41), one can derive the equation on the structure functions similar to (2.10):

$$\frac{\partial S_2}{\partial t} = -\frac{\partial S_3}{3\partial x} - 4\varepsilon + v \frac{\partial^2 S_2}{\partial x^2} . \tag{2.43}$$

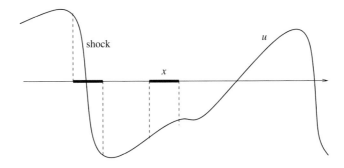

Figure 2.13 Typical velocity profile in Burgers turbulence.

Here $\varepsilon = v\langle u_x^2 \rangle$. Equation (2.43) describes both a free decay (then ε depends on t) and the case of a permanently acting pumping that generates turbulence that is statistically steady for scales less than the pumping length. In the first case, $\partial S_2/\partial t \simeq S_2 u/L \ll \varepsilon \simeq u^3/L$ (where L is a typical distance between shocks) while in the second case $\partial S_2/\partial t = 0$. In both cases, $S_3 = -12\varepsilon x + 3v\partial S_2/\partial x$. Consider now the limit $v \to 0$ at fixed x (and fixed t for decaying turbulence). Shock dissipation provides for a finite limit of ε at $v \to 0$, which gives (2.42). Similarly to incompressible turbulence, a flux constancy fixes $S_3(x)$, which is universal, determined solely by ε and depends neither on the initial statistics for decay nor on the pumping for steady turbulence. Higher moments can be related to the additional integrals of motion, $E_n = \int u^{2n}\,dx/2$, which are all formally conserved in the inviscid case. In reality, any shock dissipates the finite amount ε_n of E_n in the limit $v \to 0$ so that one can express S_{2n+1} via these dissipation rates for integer n: $S_{2n+1} \propto \varepsilon_n x$ (see Exercise 2.5). That means that the statistics of velocity differences in the inertial interval depend on the infinitely many pumping-related parameters, the fluxes of all dynamical integrals of motion.

For incompressible (vortex) turbulence described in Section 2.2.2, we have neither understanding of structures nor classification of the conservation laws responsible for an anomalous scaling.

2.3.5 Mach number

Compressibility leads to finiteness of the propagation speed of perturbations. Here we consider the motions (of the fluid or bodies) with velocity exceeding the sound velocity. The propagation of perturbations in more than one dimension is peculiar for supersonic velocities. Indeed, consider fluid moving uniformly with velocity $\mathbf{v}$. If a small disturbance appears at some place O, it will propagate with respect to the fluid with the sound velocity c. All possible velocities of propagation in the rest frame are given by $\mathbf{v} + c\mathbf{n}$ for all possible directions of the unit vector $\mathbf{n}$. This means that in a subsonic case ($v < c$) the perturbation propagates in all directions around the source O and eventually spreads to the whole fluid. This is seen from Figure 2.14, where the left circle

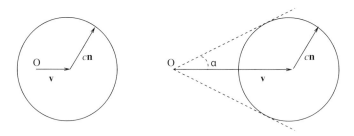

Figure 2.14 Perturbation generated at O in a fluid that moves with a subsonic (left) and supersonic (right) speed v. No perturbation can reach the outside of the Mach cone shown by broken lines.

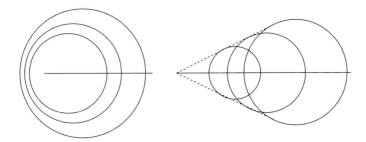

Figure 2.15 Circles are constant-phase surfaces of an acoustic perturbation generated in a fluid that moves to the right with a subsonic (left) and supersonic (right) speed. Alternatively, this may be seen as sound generated by a source moving to the left.

contains O. However, in a supersonic case, vectors $\mathbf{v} + c\mathbf{n}$ all lie within a 2α-cone with $\alpha = \arcsin c/v$ called the Mach angle. Outside the Mach cone, shown in Figure 2.14 by broken lines, the fluid stays undisturbed. The dimensionless ratio $v/c = \mathscr{M}$ is called the Mach number, which is a control parameter like Reynolds number, flows are similar for the same Re and $\mathscr{M}$.

If sound is generated, say, by periodic pulsations of a source moving relative to the fluid, the circles in Figure 2.15 correspond to the lines of a constant phase. The wavelength (the distance between the constant-phase surfaces) is smaller to the left of the source by the factor $1 - v/c$ because the propagation speed $c - v$ is smaller. For the case of a moving source, this means that the wavelength is shorter in front of the source and longer behind it. For the case of a moving fluid, it means that the wavelength is shorter upwind. The frequencies registered by the observer are, however, different in these two cases:

(i) When the emitter and receiver are at rest, the frequencies emitted and received are the same; the wavelength received upwind is smaller by a factor $1 - v/c$ and downwind larger by a factor $1 + v/c$ in a moving fluid.

(ii) When the source moves toward the receiver while the fluid is still, the propagation speed is c and the smaller wavelength corresponds to the frequency received being larger by the factor $1/(1 - v/c)$. This frequency change due to a relative motion of source and receiver is called the Döppler effect.

Döppler effect is used to determine experimentally the fluid velocity by scattering sound or light on particles carried by the flow. The effect is also used by police to catch us speeding.[16] Döppler radar emits wave with the frequency ω_0 toward a mirror (car) approaching with the speed v. In the mirror reference frame, the wave is received and reflected with the frequency $\omega_0(1 + v/c)$;

then police detector receives from a moving source the frequency, which is higher by yet another factor $1/(1-v/c)$, so that the frequency received is $\omega_0(c+v)/(c-v)$. To better appreciate that, imagine a source producing wave maxima every second so that the distance between them is c. The maxima hit the moving mirror every $c/(c+v)$ seconds and are reflected back, but every subsequent maximum meets the mirror closer to the source, so the distance between them (wavelength) after reflection is $c(c-v)/(c+v)$, and they come back to the source with the speed c every $(c-v)/(c+v)$ seconds.

Let us now describe the frequency dependence on the direction of propagation. Consider the case of the fluid which moves relative to the receiver. The latter then registers frequency that is different from the frequency ck measured in the fluid frame, where the monochromatic wave is $\exp(i\mathbf{k}\cdot\mathbf{r}' - ckt)$. The coordinates in the moving and rest frames are related as $\mathbf{r}' = \mathbf{r} - \mathbf{v}t$ so in the rest frame we have $\exp(i\mathbf{k}\cdot\mathbf{r} - ckt - \mathbf{k}\cdot\mathbf{v}t) = \exp(i\mathbf{k}\cdot\mathbf{r} - \omega_k t)$, which means that the frequency measured in the rest frame is as follows:

$$\omega_k = ck + (\mathbf{k}\cdot\mathbf{v}). \qquad (2.44)$$

This change of frequency, $(\mathbf{k}\cdot\mathbf{v})$, is called the Döppler shift. When sound propagates upwind, one has $(\mathbf{k}\cdot\mathbf{v}) < 0$, so that a standing person hears a lower tone than those downwind. Another way to put it is that the wave period is larger since more time is needed for a wavelength to pass our ear as the wind sweeps it.

Consider now a wave source that oscillates with frequency ω_0 and moves with the velocity $\mathbf{u}$. The frequency registered in the rest frame depends on the direction of propagation. Indeed, to relate ω_0 to the frequency in still air $\omega = ck$, pass to the reference frame moving with the source where $\omega_k = \omega_0$ and the fluid moves with $-\mathbf{u}$ so that (2.44) gives

$$\omega_0 = ck - (\mathbf{k}\cdot\mathbf{u}) = \omega[1 - (u/c)\cos\theta], \qquad (2.45)$$

where θ is the angle between $\mathbf{u}$ and $\mathbf{k}$. In the fluid reference frame one receives $\omega = \omega_0/[1 - (u/c)\cos\theta]$.

Let us now look at (2.44) for $v > c$. We see that the frequency of sound registered in the rest frame turns into zero on the Mach cone (also called the characteristic surface). The condition $\omega_k = 0$ defines in $\mathbf{k}$-space the cone surface $ck = -\mathbf{k}\cdot\mathbf{v}$ or in any plane the relation between the components: $v^2 k_x^2 = c^2(k_x^2 + k_y^2)$. The propagation fronts of perturbation in the x,y-plane are determined by the constant-phase condition $k_x dx + k_y dy = 0$ and $dy/dx = -k_x/k_y = \pm c/\sqrt{v^2 - c^2}$, which again corresponds to the broken straight lines in Figure 2.15 with the same Mach angle $\alpha = \arcsin(c/v) = \arctan(c/\sqrt{v^2 - c^2})$. We thus see that there is a stationary perturbation along the Mach surface, with acoustic waves inside it and undisturbed fluid outside. The Mach surface is an example of caustic, which is a general term for boundary between a region with no waves and a region with two group of waves. Indeed, two constant-phase circular surfaces intersect in every point inside the cone. We give general description of caustics in Section 3.1.4 below.

Let us now consider a flow past a body in a compressible fluid. For a slender body, like a wing, the flow perturbation can be considered small, as in Section 1.5.4, only now including density: $u + \mathbf{v}$, $\rho_0 + \rho'$, $P_0 + P'$. For small perturbations, $P' = c^2\rho'$. Linearization of the steady Euler and continuity equations gives [17]

$$\rho_0 u \frac{\partial \mathbf{v}}{\partial x} = -\nabla P' = -c^2 \nabla \rho', \qquad u \frac{\partial \rho'}{\partial x} = -\rho_0 \mathrm{div}\, \mathbf{v}. \tag{2.46}$$

Taking the curl of the Euler equation, we get $\partial \omega / \partial x = 0$. Since vorticity is x-independent and zero far upstream, it is zero everywhere in a linear approximation.[18] We thus have a potential flow, $\mathbf{v} = \nabla \phi$, which satisfies

$$\left(1 - \mathscr{M}^2\right)\frac{\partial^2 \phi}{\partial x^2} + \frac{\partial^2 \phi}{\partial y^2} = 0. \tag{2.47}$$

Here the Mach number, $\mathscr{M} = u/c$, determines whether the equation is elliptic or hyperbolic. When $\mathscr{M} < 1$, the equation is elliptic and the streamlines are everywhere smooth. When $\mathscr{M} > 1$, the equation is hyperbolic and the streamlines have cusps on the Mach planes, extending from the ends of the wing; the streamlines are straight outside and curved only between the planes, see Figure 2.16.

In the elliptic case, the change of variables $x \to x(1 - \mathscr{M}^2)^{-1/2}$ turns (2.47) into a Laplace equation, $\mathrm{div}\, \mathbf{v} = \Delta \phi = 0$, which we had for an incompressible case. To put it simply, at subsonic speeds, compressibility of the fluid is equivalent to a longer body. Since the lift is proportional to the velocity circulation, i.e. to the wing length, then we conclude that compressibility increases the lift by $(1 - \mathscr{M}^2)^{-1/2}$.

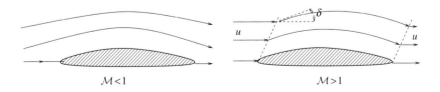

Figure 2.16 Subsonic (left) and supersonic (right) flow around a slender wing.

In the hyperbolic case, the solution is

$$\phi = F(x - By) , \quad B = (\mathcal{M}^2 - 1)^{-1/2} .$$

The boundary condition on the body having the shape $y = f(x)$ is $v_y = \partial\phi/\partial y = uf'(x)$, which gives $F = -uf/B$. This means that $v_y/u = f'(x)$ everywhere, that is the streamlines repeat the body shape and turn straight behind the rear Mach surface (in the linear approximation). We see that passing through the Mach surface the velocity and density have a jump proportional to $f'(0)$. This means that Mach surfaces (like the planes or cones described here) are actually shocks. One can relate the flow properties before and after the shock by the conservation laws of mass, energy and momentum, in this case called the Rankine–Hugoniot relations. Namely, if w is the velocity component normal to the front, then the fluxes ρw, $\rho w(W + w^2/2) = \rho w[\gamma P/(\gamma - 1)\rho + w^2/2]$ and $P + \rho w^2$ must be continuous through the shock. This gives three relations that can be solved for the pressure, velocity and density after the shock (Exercise 2.3). In particular, for a slender body when the streamlines deflect by a small angle $\delta = f'(0)$ after passing through the shock, we get the velocity changes $v_y = \partial\phi/\partial y = u\delta$ and $v_x = \partial\phi/\partial x = -u\delta/B$. The pressure change due to the velocity decrease is as follows:

$$\frac{\Delta P}{P} \propto \frac{u^2 - (u + v_x)^2 - v_y^2}{c^2} = \mathcal{M}^2 \left[1 - \left(1 - \frac{\delta}{\sqrt{\mathcal{M}^2 - 1}} \right)^2 - \delta^2 \right]$$

$$\approx \frac{2\delta \mathcal{M}^2}{\sqrt{\mathcal{M}^2 - 1}} . \tag{2.48}$$

The compressibility contribution to the drag is proportional to the pressure drop and thus the drag jumps when $\mathcal{M}$ crosses unity, owing to the appearance of shock and the loss of acoustic energy radiated away between the Mach planes. The drag and lift singularity at $\mathcal{M} \to 1$ is sometimes referred to as the "sound barrier." Apparently, our assumption of small perturbations does not work at $\mathcal{M} \to 1$. For comparison, recall that the wake contribution to the drag is proportional to ρu^2, while the shock contribution (2.48) is proportional to $P\mathcal{M}^2/\sqrt{\mathcal{M}^2 - 1} \simeq \rho u^2/\sqrt{\mathcal{M}^2 - 1}$.

We see that in a linear approximation, the steady-state two-dimensional flow perturbation does not decay with distance. We have learnt in Section 2.3.2 that the propagation speed depends on the amplitude and so must the angle α, which means that the Mach surfaces are straight only where the amplitude is small, which is usually far away from the body. Weak shocks with $\mathcal{M} - 1 \ll 1$ can be described by the Burgers equation. Indeed, according to (2.39) and (2.40), the front width is

$$\frac{v}{u-c} = \frac{v}{c(\mathcal{M}-1)} \simeq \frac{lv_T}{c(\mathcal{M}-1)},$$

which exceeds the mean free path l only for $\mathcal{M}-1 \ll 1$ since the molecular thermal velocity v_T and the sound velocity c are comparable (see also Section 1.4.4). To be consistent in the framework of continuous media, strong shocks must be considered as discontinuities.

Exercises

2.1 (i) Two-dimensional incompressible flow around a saddle-point corresponds to a pure strain: $v_x = \lambda x$, $v_y = -\lambda y$. The coordinates $x(t), y(t)$ of a fluid particle satisfy the equations $\dot{x} = v_x$ and $\dot{y} = v_y$. Whether the vector $\mathbf{r} = (x, y)$ is stretched or contracted after some time T depends on its orientation and on T. Find which fraction of the vectors is stretched.

(ii) Consider a two-dimensional incompressible flow having both permanent strain λ and vorticity ω: $v_x = \lambda x + \omega y/2$, $v_y = -\lambda y - \omega x/2$. Describe the motion of the particle, $x(t), y(t)$, for different relations between λ and ω.

2.2 Consider a fluid layer between two horizontal parallel plates kept at the distance h at temperatures that differ by Θ. The fluid has kinematic viscosity v, thermal conductivity χ (both measured in cm^2 s^{-1}) and the coefficient of thermal expansion $\beta = -\partial \ln \rho / \partial T$, such that the relative density change due to the temperature difference, $\beta\Theta$, far exceeds the change due to the hydrostatic pressure difference, gh/c^2, where c is the velocity of sound. Find the control parameter(s) for the appearance of the convective (Rayleigh–Bénard) instability.

2.3 Taylor dispersion along a narrow pipe. Derive the laminar Poiseuille profile $v(r)$ carrying mean flux $\bar{v}$ in a circular pipe of radius a. Substitute this velocity profile into the advection-diffusion equation (2.17) and derive the equation on the concentration averaged over the cross-section, $\bar{\theta}(x,t) = \int \theta(x, \mathbf{r}, t) \, d\mathbf{r}/\pi a^2$, considering times exceeding a^2/κ, were κ is the molecular diffusivity. Find the effective diffusivity along the pipe.

2.4 Consider a shock wave with the velocity w_1 normal to the front in a polytropic gas having the enthalpy

$$W = c_p T = PV \frac{\gamma}{\gamma-1} = \frac{P}{\rho}\frac{\gamma}{\gamma-1} = \frac{c^2}{\gamma-1},$$

where $\gamma = c_p/c_v$. Write Rankine–Hugoniot relations for this case. Express the ratio of densities ρ_2/ρ_1 via the pressure ratio P_2/P_1, where the subscripts 1 and 2 denote the values before and after the shock. Express P_2/P_1, ρ_2/ρ_1 and $\mathcal{M}_2 = w_2/c_2$ via the preshock Mach number $\mathcal{M}_1 = w_1/c_1$. Consider the limits of strong and weak shocks.

2.5 For Burgers turbulence, express the fifth structure function S_5 via the dissipation rate $\varepsilon_4 = 6\nu[\langle u^2 u_x^2 \rangle + \langle u^2 \rangle \langle u_x^2 \rangle]$.

2.6 In a standing sound wave, air with density ρ locally moves as follows: $v = a\sin(\omega t)$. Consider a small suspended spherical particle with density ρ_0 whose material evaporates so that its volume $V(t)$ decreases with the rate proportional to the relative velocity: $V^{-1}dV/dt = -\alpha|v - u|$. Find how the particle velocity u changes with time if it was initially at rest. Assume $\rho_0 \gg \rho$.

2.7 There is anecdotal evidence that early missiles suffered from an interesting malfunction of the fuel gauge. The gauge was a simple floater (a small air-filled rubber balloon) whose position was supposed to signal the level of liquid fuel during the ascending stage. However, when the engine was warming up before starting, the gauge unexpectedly sank to the bottom, signalling zero level of fuel and shutting off the engine. How do engine-reduced vibrations reverse the sign of effective gravity for the balloon in the fluid?

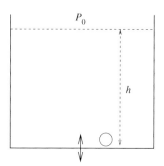

Consider an air bubble in the vessel filled up to the depth h by a liquid with density ρ. The vessel vibrates vertically according to $x(t) = (Ag/\omega^2)\sin(\omega t)$, where g is the static gravity acceleration. Find the threshold amplitude A necessary to keep the bubble near the bottom. The pressure on the free surface is P_0.

2.8 It is a common experience that acoustic intensity drops quickly with the distance when the sound propagates upwind. Why is it so difficult to hear somebody shouting against the wind?

2.9 Describe an axial symmetric propagation of acoustic waves. Hint: use (2.27).

2.10 The relation (2.45), $\omega_0 = \omega[1 - (u/c)\cos\theta]$, suggests that the frequencies ω_0 (emitted) and ω (received) have different signs when a sound source moves toward a receiver with a supersonic speed, $u\cos\theta > c$. What does it mean physically?

2.11 Consider a steady flow of a compressible ideal fluid with no external body force. Does the flux ρv along the streamline increase or decrease with the velocity v? Hint: express pressure variation as $dp = c^2 d\rho$.

2.12 Consider a spherically symmetric radial steady flow of an isothermal ideal gas in the gravity field of a star with mass M (a model of stellar wind). Can the velocity of such a flow grow with the distance?

3

Dispersive waves

In this chapter, we consider systems that support small-amplitude waves whose speed depends on wavelength. This is in distinction from acoustic waves (or light in the vacuum) that all move with the same speed so that a small-amplitude one-dimensional perturbation propagates without changing its shape. When the speeds of different Fourier harmonics are different, the shape of a perturbation generally changes as it propagates. In particular, initially localized perturbation spreads. That is, dispersion of wave speed leads to packet dispersion in space. This is why such waves are called dispersive. Since different harmonics move with different speeds, then they separate with time and can subsequently be found in different places. As a result, for quite arbitrary excitation mechanisms one often finds locally sinusoidal perturbation, the property well known to everybody who has observed waves on water surface. We often see periodic water waves and rarely hear pure tones because surface waves are dispersive while sound waves are not. Surface waves form the main subject of analysis in this section but the ideas and results apply equally well to numerous other dispersive waves that exist in bulk fluids, plasma and solids. Dispersion usually results from some anisotropy or inhomogeneity of the medium. We shall try to keep our description universal when we turn to a consideration of weakly nonlinear waves having small but finite amplitudes. In this case, we also consider dispersion weak, which is possible in two distinct cases: (i) when the dispersion relation is close to acoustic and (ii) when waves are excited in a narrow spectral interval. These two cases correspond respectively to the Korteveg–de-Vries equation and the nonlinear Schrödinger equation, which are as universal for dispersive waves as the Burgers equation for nondispersive waves. In particular, the results of this chapter are as applicable to nonlinear optics and quantum physics as to fluid mechanics. We have seen that without dispersion nonlinearity leads to breaking of sound waves and creation of shocks. Here we shall see that nonlinear steepening and dispersive spreading can balance each other to create stationary nonlinear waves, in particular, soliton, quintessentially nonlinear object.

3.1 Linear waves

To have waves, one either needs inhomogeneity of fluid properties or compressibility. Here we consider an incompressible fluid with an extreme form of inhomogeneity – an open surface. In this section, we shall consider surface waves as an example of dispersive wave systems, account for gravity and surface tension as restoring forces and describe the effects of viscous friction. We then introduce general notions of phase and group velocities, which are different for dispersive waves. We discuss physical phenomena that appear because of that difference.

Linear waves in an infinite medium can be presented as superpositions of plane waves $\exp(i\mathbf{kr} - i\omega t)$. Therefore, linear waves of every type are completely characterized by the so-called dispersion relation between wave frequency ω and wavelength $\lambda = 2\pi/k$. As befits physicists and engineers, before making formal derivations we show simple ways to estimate $\omega(\lambda)$ up to a numerical factor. The simplest way is usually dimensional analysis. If gravity is a restoring force then the only relation between ω, λ, g is $\omega^2 \simeq g\lambda^{-1}$. If surface tension dominates, the frequency must depend on the coefficient of surface tension α (having dimensionality force/length = gram/s^2) as well as fluid density ρ which characterizes inertia. In this case, we have four parameters, $\omega, \lambda, \alpha, \rho$, and three dimensionalities, *gram, centimetre and second*, so that according to the π-theorem from Section 1.4.4 the expression for the frequency is unique again: $\omega^2 \simeq \alpha\lambda^{-3}\rho^{-1}$ up to a dimensionless factor.

If, however, gravity and surface tension are comparable, then five parameters and three dimensionalities does not allow one to determine the dispersion relation from dimensional analysis. One then ascends to a bit higher (yet still elementary) level of making an estimate by using Newton's second law or its equivalent for small oscillations, called the virial theorem, which states that mean kinetic energy is equal to the mean potential energy.

Consider vertical oscillations of a fluid with the elevation amplitude a and the frequency ω. The fluid velocity can be estimated as ωa and the acceleration as $\omega^2 a$. When the fluid depth is much larger than the wavelength, we may assume that the fluid layer of the order λ is involved in the motion. Newton's second law for a unit area then requires that the mass $\rho\lambda$ times the acceleration $\omega^2 a$ must be equal to the gravitational force $\rho a g$:

$$\rho\omega^2 a\lambda \simeq \rho a g \Rightarrow \omega^2 \simeq g\lambda^{-1}. \tag{3.1}$$

The same result one obtains using the virial theorem. Indeed, the kinetic energy per unit area of the surface can be estimated as the mass in motion $\rho\lambda$ times the velocity squared $\omega^2 a^2$. The gravitational potential energy per unit area can be estimated as the elevated mass ρa times g times the elevation a.

A curved surface has extra potential energy, whose density per unit area is the product of the coefficient of surface tension α and the surface curvature $(a/\lambda)^2$. Taking potential energy as a sum of gravitational and capillary contributions, we obtain the dispersion relation for gravitational-capillary waves on deep water:

$$\omega^2 \simeq g\lambda^{-1} + \alpha\lambda^{-3}\rho^{-1}. \tag{3.2}$$

When the depth h is much smaller than λ, fluid mostly moves horizontally with the horizontal velocity exceeding the vertical velocity ωa by the geometric factor λ/h. The mass that moves is now ρh. This makes the kinetic energy $\rho\omega^2 a^2\lambda^2 h^{-1}$ while the potential energy does not change. The virial theorem then gives the dispersion relation for waves on shallow water:

$$\omega^2 \simeq gh\lambda^{-2} + \alpha h\lambda^{-4}\rho^{-1}. \tag{3.3}$$

3.1.1 Surface gravity waves

Let us now formally describe fluid motion in a surface wave. As argued in Section 1.2.4, small-amplitude oscillations are irrotational flows. Then one can introduce the velocity potential that satisfies the Laplace equation $\Delta\phi = 0$ for incompressible flows. The pressure is

$$p = -\rho(\partial\phi/\partial t + gz + v^2/2) \approx -\rho(\partial\phi/\partial t + gz),$$

neglecting quadratic terms because the amplitude is small. As in considering Kelvin–Helmholtz instability in Section 2.1 we describe the surface form by the elevation $\zeta(x,t)$. We can include the atmospheric pressure on the surface into $\phi \to \phi + p_{0t}/\rho$, which does not change the velocity field. We then have on the surface

$$\frac{\partial\zeta}{\partial t} = v_z = \frac{\partial\phi}{\partial z}, \quad g\zeta + \frac{\partial\phi}{\partial t} = 0 \quad \Rightarrow \quad g\frac{\partial\phi}{\partial z} + \frac{\partial^2\phi}{\partial t^2} = 0. \tag{3.4}$$

The first equation here is the linearized kinematic boundary condition (2.1), which states that the vertical velocity on the surface is equal to the time derivative of the surface height. The second one is the linearized dynamic boundary condition on the pressure being constant on the surface; it can also be obtained writing the equation for the horizontal acceleration due to gravity acting on an inclined surface: $dv_x/dt = -g\partial\zeta/\partial x$. To solve (3.4) together with $\Delta\phi = 0$, one needs the boundary condition at the bottom: $\partial\phi/\partial z = 0$ at $z = -h$. The solution of the Laplace equation periodic horizontally is exponential vertically:

$$\phi(x,z,t) = a\cos(kx - \omega t)\cosh[k(z+h)], \tag{3.5}$$
$$\omega^2 = gk\tanh kh. \tag{3.6}$$

Differentiating the potential with respect to time and coordinates, we obtain

$$\zeta = -a(\omega/g)\sin(kx - \omega t)\cosh[k(z+h)], \tag{3.7}$$
$$v_x = -ak\sin(kx - \omega t)\cosh[k(z+h)], \tag{3.8}$$
$$v_z = ak\cos(kx - \omega t)\sinh[k(z+h)]. \tag{3.9}$$

Note that in the linear approximation $v_x = gk\zeta/\omega$, i.e. fluid moves forward near crests and backward near troughs, as is known to every swimmer. The condition of weak nonlinearity is $\partial v/\partial t \gg v\partial v/\partial x$, which requires $\omega \gg ak^2 = kv$ that can be written as $g \gg kv^2$.

The trajectories of fluid particles can be obtained by integrating the equation $\dot{\mathbf{r}} = \mathbf{v}$ assuming small oscillations near some $\mathbf{r_0} = (x_0, z_0)$, as for solving (2.29). Fluid displacement during the period can be estimated as velocity ka times $2\pi/\omega$; this is supposed to be much smaller than the wavelength $2\pi/k$. At first order in the small parameter ak^2/ω, we find

$$x = x_0 - \frac{ak}{\omega}\cos(kx_0 - \omega t)\cosh[k(z_0+h)], \tag{3.10}$$
$$z = z_0 - \frac{ak}{\omega}\sin(kx_0 - \omega t)\sinh[k(z_0+h)].$$

The trajectories are ellipses described by

$$\left(\frac{x - x_0}{\cosh[k(z_0+h)]}\right)^2 + \left(\frac{z - z_0}{\sinh[k(z_0+h)]}\right)^2 = \left(\frac{ak}{\omega}\right)^2.$$

We see that as one goes down away from the surface, the amplitude of the oscillations decreases and the ellipses become more elongated as one approaches the bottom.

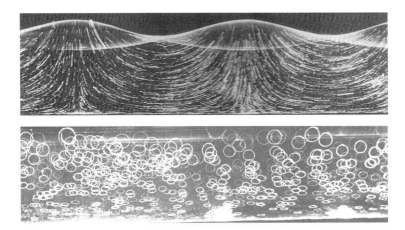

Figure 3.1 White particles suspended in the water are photographed during one period. The top figure shows a standing wave, where the particle trajectories are streamlines. The bottom figure shows a wave propagating to the right, some open loops indicate a Stokes drift to the right near the surface and compensating reflux to the left near the bottom. In both cases, the wave amplitude is 4% and the depth is 22% of the wavelength. From A. Wallett and F. Ruellan, *La Houille Blanche*, **5**, 483–489 (1950).

One can distinguish two limits, depending on the ratio of the water depth to the wavelength. On shallow water ($kh \ll 1$), the oscillations are almost one-dimensional: $v_z/v_x \propto kh$. The dispersion relation is sound-like: $\omega = \sqrt{gh}\,k$ (this formula is all one needs to answer the question in Exercise 3.1).

For gravity waves on deep water ($kh \gg 1$), the frequency, $\omega = \sqrt{gk}$, is like the formula for the period of the pendulum $T = 2\pi/\omega = 2\pi\sqrt{L/g}$. Indeed, a standing surface wave is like a pendulum made of water, as seen in the top panel of Figure 3.1. For a running wave we have

$$\zeta = -\frac{u}{\omega}\sin(kx - \omega t)\mathrm{e}^{kz}, \qquad (3.11)$$

$$v_x = -u\sin(kx - \omega t)\mathrm{e}^{kz}, \qquad (3.12)$$

$$v_z = u\cos(kx - \omega t)\mathrm{e}^{kz}. \qquad (3.13)$$

The fluid particles move in perfect circles whose radius exponentially decays with depth, with the rate equal to the horizontal wavenumber. This is again the property of the Laplace equation mentioned in Section 2.2.3: if the solution oscillates in one direction, it decays exponentially in the transverse direction. This supports our assumption that for a deep fluid, the layer comparable to wavelength is involved in motion, a fact known to divers.

At second order, we expect nonzero momentum and the mean (Stokes) drift. The reason is that the velocity potential must be purely periodic for a deep water, because the only other solution of the Laplace equation, $xf(z)$, is ruled out by the requirement $df(z)/dz \to 0$ as $z \to -\infty$. Since velocity cannot be x-independent, then counterflow to the Stokes drift is impossible, contrary to the case of sound discussed at the end of Section 2.3.1. Therefore, propagation of potential gravity waves on a deep water is always accompanied by a surface current.[1] Since the total momentum of any horizontal level inside the fluid is zero for a periodic potential, the only contribution is from surface disturbance. The nonzero momentum appears in the second order as a product of v, ζ disturbances of the first order. Multiplying (3.11) by (3.12) and averaging over the period we obtain the mean momentum density per unit area: $\rho\langle \zeta v_x \rangle = \rho u^2/2\omega$. Exactly like in (2.30), we can also obtain the drift velocity:

$$\langle v_x \rangle = -ku\langle (x-x_0)\cos(kx-\omega t) + (z-z_0)\sin(kx-\omega t) \rangle = \frac{ku^2}{\omega}. \qquad (3.14)$$

From the general expression (1.46) for the displacement, one can also obtain the drift velocity as an average of $k|\nabla\phi|^2/\omega = kv^2/\omega$ over the period. For small-amplitude wave, velocity is constant on a particle orbit and we obtain (3.14). In a deep tank of a finite length we expect the surface drift to be balanced by the counterflow at the bottom, as seen in the bottom panel of Figure 3.1. When the water is not deep one cannot disentangle surface and bottom currents; shallow-water waves can have zero mass flow, exactly like sound.

3.1.2 Viscous dissipation

The moment we account for viscosity, our solution (3.5) does not satisfy the boundary condition on the free surface,

$$\sigma_{ij}n_j = \sigma'_{ij}n_j - pn_i = 0$$

(see Section 1.4.3), because both the tangential stress $\sigma_{xz} = -2\eta\phi_{xz}$ and the oscillating part of the normal stress $\sigma'_{zz} = -2\eta\phi_{zz}$ are nonzero. Note also that (3.12) gives v_x nonzero on the bottom. A true viscous solution has to be rotational but when viscosity is small, vorticity appears only in narrow boundary layers near the surface and the bottom. A standard derivation[2] of the rate of a weak decay is to calculate the viscous stresses from the solution (3.5) and substitute them into (1.52):

$$\frac{dE}{dt} = -\frac{\eta}{2}\int \left(\frac{\partial v_i}{\partial x_k} + \frac{\partial v_k}{\partial x_i}\right)^2 dV = -2\eta\int \left(\frac{\partial^2\phi}{\partial x_k\partial x_i}\right)^2 dV$$

$$= -2\eta\int \left(\phi_{zz}^2 + \phi_{xx}^2 + 2\phi_{xz}^2\right) dV = -8\eta\int \phi_{xz}^2\, dV.$$

This means neglecting the narrow viscous boundary layers near the surface and the bottom (where there is not much motion by virtue of $kh \gg 1$). Assuming the decay to be weak (this requires $\nu k^2 \ll \omega$, which also guarantees that the boundary layer width $\sqrt{\nu/\omega}$ is smaller than the wavelength), we consider the energy averaged over the period, which is twice the averaged kinetic energy for small oscillations by the virial theorem:

$$\bar{E} = \int_0^{2\pi/\omega} E \, dt \, \omega/2\pi = \rho \int \overline{v^2} \, dV = 2\rho k^2 \int \overline{\phi^2} \, dV.$$

The dissipation averaged over the period is related to the average energy:

$$\frac{\overline{dE}}{dt} = \int_0^{2\pi/\omega} \frac{dE}{dt} \, dt \, \omega/2\pi = -8\eta k^4 \int \overline{\phi^2} \, dV = -4\nu k^2 \bar{E}. \qquad (3.15)$$

If you are not confused, you are not paying attention. Indeed, something seems strange in this derivation. Notice that our solution (3.5) satisfies the Navier–Stokes equation (but not the boundary conditions) since the viscous term is zero for the potential flow: $\Delta \mathbf{v} = 0$. How then can zero viscous force give nonzero dissipation? To answer that, note that the force $\rho \nu \Delta \mathbf{v} = \partial \sigma'_{ik}/\partial x_k$ is the viscous stress divergence, which is zero. But the stress σ'_{ik} itself is nonzero. In other words, the *net* viscous force on any fluid element is zero, but the viscous forces around an element are nonzero as it deforms (compare with Exercise 1.16). These forces bring the energy dissipation, which is $\sigma'_{ik} \partial v_i/\partial x_k$. Indeed, we have shown in (1.51) that the viscous contribution to the energy change consists of two terms: $v_i \partial \sigma'_{ik}/\partial x_k = \partial[v_i \sigma'_{ik}]/\partial x_k - \sigma'_{ik} \partial v_i/\partial x_k$. The first term is the divergence and describes the transport of energy, while the second term describes dissipation. This second term has a nonzero time average so that viscosity causes waves to decay. Moreover, $\sigma'_{ik} \partial v_i/\partial x_k$ is distributed over the fluid bulk rather than being concentrated in the viscous boundary layer, in distinction from the term $v_i \partial \sigma'_{ik}/\partial x_k = \mathbf{v} \Delta \mathbf{v}$. To appreciate how the bulk integration can present the boundary layer distortion, consider a function $U(x)$ on $x \in [0, 1]$, which is almost linear but curves in a narrow vicinity near $x = 1$ to give $U'(1) = 0$, as shown in Figure 3.2. Then the integral $\int_0^1 UU'' \, dx = -\int_0^1 (U')^2 \, dx$ is nonzero, and the main contribution is from the bulk.

And yet the first derivation of the viscous dissipation (3.15) was made by Stokes in quite an ingenious way, which did not involve any boundary layers. Since the potential solution satisfies the Navier–Stokes equation, he suggested *imagining* how one may also satisfy the boundary conditions (so that no boundary layers appear). First, we need an extensible bottom to move with $v_x(-h) = -k\sin(kx - \omega t)$. Since $\sigma'_{xz}(-h) = 0$, such bottom movements do no work. In addition, we must apply extra forces to the fluid surface to

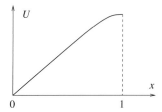

Figure 3.2 A function whose U'' is small everywhere except a small vicinity of $x = 1$ yet $\int_0^1 UU'' \, dx = - \int_0^1 (U')^2 \, dx$ is due to the bulk.

compensate for $\sigma_{zz}(0)$ and $\sigma_{xz}(0)$. Such forces do work (per unit area per unit time) $v_x \sigma_{xz} + v_z \sigma_{zz} = 2\eta (\phi_x \phi_{xz} + \phi_z \phi_{zz})$. After averaging over the period of the monochromatic wave, this becomes $4\eta k^2 \overline{\phi \phi_z}$, which is $8\nu k^2$ times the average kinetic energy per unit surface area, $\rho \overline{\phi \phi_z}/2$ – to obtain this, one writes

$$\rho \int (\nabla \phi)^2 \, dV = \rho \int \nabla \cdot (\phi \nabla \phi) \, dV = \int \phi \phi_z \, dS.$$

Since we have introduced forces that make our solution steady (it satisfies equations and boundary conditions), the work of those forces exactly equals the rate of the bulk viscous dissipation which is thus $4\nu k^2$ times the total energy.[3]

3.1.3 Capillary waves

Surface tension creates an extra pressure difference proportional to the curvature of the surface:

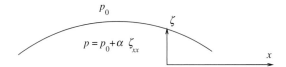

This changes the second equation of (3.4),

$$g\zeta + \frac{\partial \phi}{\partial t} = \frac{\alpha}{\rho} \frac{\partial^2 \zeta}{\partial x^2}, \tag{3.16}$$

and adds an extra term to the dispersion relation

$$\omega^2 = \left(gk + \alpha k^3 / \rho \right) \tanh kh. \tag{3.17}$$

Indeed, the two restoring factors enter additively into ω^2, as was suggested by the virial theoremat in the beginning of Section 3.1.

We see that there is a wavenumber $k_* = \sqrt{\rho g / \alpha}$, which separates gravity-dominated long waves from short waves dominated by surface tension. In general, both gravity and surface tension provide a restoring force for the surface perturbations. For water, $\alpha \simeq 70$ erg cm^{-2} and $\lambda_* = 2\pi / k_* \simeq 1.6$ cm. Now we can answer why water cannot be held in an upside-down glass. The question seems bizarre only to a nonphysicist, since physicists usually know that atmospheric pressure at normal conditions is of the order of $P_0 \simeq 10^5$ newtons per square metre, which is enough to support up to $P_0 / \rho g \simeq 10$ metres of water column.[4] That is, if the fluid surface remained plane then the atmospheric pressure would keep the water from spilling. The relation (3.17) tells us that with a negative gravity, the plane surface is unstable:

$$\omega^2 \propto -gk + \alpha k^3 < 0 \quad \text{for} \quad k < k_*.$$

On the contrary, water can be kept in an upside-down capillary with a diameter smaller than λ_* since the unstable modes cannot fit in (see Exercise 3.10 for the consideration of a more general case).

3.1.4 Phase and group velocity

Let us discuss now general properties of one-dimensional propagation of linear dispersive waves. To describe the propagation, we use the Fourier representation, since every harmonic $\exp(ikx - i\omega_k t)$ propagates with the constant velocity ω_k / k, completely determined by the frequency–wavenumber relation ω_k. One can learn quite general lessons by considering the perturbation in the simple Gaussian form,

$$\zeta(x,0) = \zeta_0 \exp\left(ik_0 x - x^2 / l^2\right),$$

for which the distribution in k-space is Gaussian as well:

$$\zeta(k,0) = \int dx \zeta_0 \exp\left[i(k_0 - k)x - \frac{x^2}{l^2}\right] = \zeta_0 l \sqrt{\pi} \exp\left[-\frac{l^2}{4}(k_0 - k)^2\right].$$

This distribution has the width $1/l$. Consider first a spatially wide and spectrally narrow wave packet (quasimonochromatic wave) with $k_0 l \gg 1$ so that we can expand

$$\omega(k) = \omega_0 + (k - k_0)\omega' + (k - k_0)^2 \omega'' / 2 \qquad (3.18)$$

and substitute it into

$$\zeta(x,t) = \int \frac{dk}{2\pi} \zeta(k,0) \exp[ikx - i\omega_k t] \qquad (3.19)$$

$$\approx \frac{\zeta_0 l}{2} e^{ik_0 x - i\omega_0 t} \left[l^2 / 4 + it\omega'' / 2\right]^{-1/2} \exp\left[-\frac{(x - \omega' t)^2}{(l^2 + 2it\omega'')}\right]. \qquad (3.20)$$

We see that the perturbation $\zeta(x,t) = \exp(ik_0 x - i\omega_0 t)\Psi(x,t)$ is a monochromatic wave with a complex envelope having the modulus

$$|\Psi(x,t)| \approx \zeta_0 \frac{l}{L(t)} \exp\left[-(x-\omega' t)^2/L^2(t)\right],$$

$$L(t) = [l^4 + (t\omega''/2)^2]^{1/4}. \tag{3.21}$$

The phase propagates with the *phase velocity* ω_0/k_0, while the envelope (and the energy determined by $|\Psi|^2$) propagate with the *group velocity*, ω'. The maximum is at $x = \omega' t$ where waves with close wavevectors interfere constructively, away from this point the waves cancel each other. For sound waves, $\omega_k = ck$, the group and the phase velocities are equal and are the same for all wavenumbers; the wave packet does not spread, since $\omega'' = 0$. On the contrary, for dispersive waves with $\omega'' \neq 0$, the wave packet spreads with time and its amplitude decreases since $L(t)$ grows. For $\omega_k \propto k^\alpha$, $\omega' = \alpha \omega_k/k$. In particular, the group velocity for gravity waves on deep water is half that of the phase velocity so that individual crests can be seen appearing out of nowhere at the back of the packet and disappearing at the front. For capillary waves on deep water, the group velocity is 1.5 times more, so crests appear at the front.

Consider now a spatially localized initial perturbation, which corresponds to a wide distribution in k-space with many harmonics coordinating their phases in such a way as to provide constructive interference inside a narrow region and destructive interference outside. For dispersive waves ($\omega'' \neq 0$) different harmonics propagate with different velocities and separate. After a long time, we shall see periodic perturbation with the wavelength depending on the position. Indeed, in the integral (3.19) for given large x,t the main contribution is given by the wavenumber determined by the stationary phase condition $\omega'(k_c) = x/t$, i.e. waves of wavenumber $k_c(x,t)$ are found at positions moving forward with the group velocity $\omega'(k_c)$. The spectral form of the perturbation is irrelevant in this limit and can be considered constant. Substituting $\zeta(k,0) = 1$ (or taking a limit $l \to 0$, $\zeta_0 l/\sqrt{\pi} = 1$) we obtain from (3.19):

$$\zeta(x,t) \approx \frac{1}{2} e^{ik_c x - i\omega(k_c)t} \left[it\omega''(k_c)/2\right]^{-1/2}. \tag{3.22}$$

The dependence of the wave envelope on x,t is thus determined by the factor $[t\omega''(k_c)]^{-1/2}$. Water waves have nonmonotonic dependence of the group velocity on the wavenumber so that the equation $\omega'(k_c) = x/t$ has two solutions. Smaller wavenumber corresponds to a gravity wave and larger wavenumber to a capillary wave; these waves then can propagate together, see Exercise 3.3. For gravity waves, $\omega'(k_c) = \sqrt{g/4k_c} = x/t$ gives $k_c = gt^2/4x^2$ and the envelope behaves as $|\zeta(x,t)| \propto \sqrt{gt^2/x^3}$. It decays with distance at a given time and grows with time at a given point. For capillary waves, $\omega'(k_c) = (3/2)\sqrt{\alpha k_c/\rho} = x/t$ gives $k_c = 4\rho x^2/9\alpha t^2$ and $|\zeta(x,t)| \propto x^{1/2}/t$. Counterintuitively, we see that the amplitude of the capillary wave train actually *grows* with the distance. That growth is restricted by the condition $k_c l \ll 1$ which restricts the distance by $x \ll \sqrt{\alpha/\rho l}$. At larger distances, the amplitude decays with the distance; in reality, short capillary waves are effectively attenuated by viscosity.

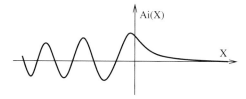

The consideration leading to (3.22) apparently breaks down as we approach the point where $k_c(x,t) = k_*$, such that $\omega''(k_*) = 0$, that is where the group velocity has maximum or minimum. That point is caustic since it is the boundary between a region with no waves and a region with two group of waves. Group velocity is stationary on caustics so that waves with close wavevectors run together. Describing the long-time behavior at the vicinity of the caustic requires further expansion of $\omega(k)$ up to cubic terms. Assuming $\omega'(k_*) - x/t$ small but nonzero, denoting $\omega(k_*) = \omega_*$ and $\omega'(k_*) = v_*$, we obtain

$$\zeta(x,t) \approx e^{ik_*x - i\omega_*t} \int \frac{dk}{2\pi} \exp[i(k-k_*)(x-v_*t) - i(k-k_*)^3 \omega'''(k_*)t/6]$$

(3.23)

$$= \frac{e^{ik_*x - i\omega_*t}}{2\pi[\omega'''(k_*)t/2]^{1/3}} \int ds\, e^{isX + is^3/3} = \frac{e^{ik_*x - i\omega_*t}}{[\omega'''(k_*)t/2]^{1/3}} Ai(X).$$

Here we denoted $X = (v_*t - x)[\omega'''(k_*)t/2]^{-1/3}$ and introduced the Airy integral $Ai(X)$ playing near caustic the same role as the Gaussian integral (3.20) at a general point. The Airy integral as a function of X is shown in the Figure, which can be understood, again, with the help of the stationary phase approach. Far behind the caustic where $X > 1$, the integral is determined by the imaginary saddle point $s = i\sqrt{X}$ and is exponentially small: $Ai(X) \propto \exp(-2X^{3/2}/3)$. In front of the caustics where X is negative, we have two real saddle points with $s = \pm\sqrt{X}$ and $Ai(X) \propto |X|^{-1/4} \cos(2X^{3/2}/3 - \pi/4)$. The wave amplitude reaches its maximum slightly ahead of the caustic. Note that the amplitude around the caustic decays as $t^{-1/3}$, that is slower than the decay $t^{-1/2}$ described by (3.21) for a plane quasi-monochromatic wave with a generic wavenumber.

In a d-dimensional case, the approach (3.20) based on the Gaussian integral gives instead of (3.22) the following expression

$$\zeta(\mathbf{x},t) = e^{i\mathbf{k}_c\cdot\mathbf{x}-i\omega(k_c)t} \int dk_1 \ldots dk_d \exp(-it k_i k_j \partial^2 \omega/\partial k_i \partial k_j/2)$$
$$\approx e^{i\mathbf{k}_c\cdot\mathbf{x}-i\omega(k_c)t} \left(\det[it\partial^2 \omega/\partial k_i \partial k_j/2\pi]\right)^{-1/2}. \tag{3.24}$$

The caustic appears for such $\mathbf{x},t$ where the determinant of the matrix $\partial^2 \omega/\partial k_i \partial k_j$ vanishes at $\mathbf{k}_c$ defined by $\partial \omega(k_c)/\partial \mathbf{k} = \mathbf{x}/t$. Sufficient condition for that is vanishing of a single eigenvalue of the matrix. The direction of the respective eigenvector defines the normal to the caustic, which is a $(d-1)$-dimensional surface in $\mathbf{x}$-space. We then have Airy integral along this direction and a usual Gaussian integration along remaining $d-1$ directions, which gives the amplitude on the caustic decaying as $t^{-1/3-(d-1)/2}$.

3.1.5 Wave generation

An obstacle to the stream or wave source moving with respect to water can generate a steady wave pattern if the projection of the source's relative velocity V in the direction of wave propagation is equal to the phase velocity $c(k) = \omega(k)/k$. For example, to create an elevation of the water surface the source must stay on the wave crest, which moves with the phase velocity. If the wave propagates at the angle θ to the direction of the source motion, then the condition $V\cos\theta = c$ for generation of a stationary wave pattern is a direct analogue of the resonance condition for Vavilov–Cherenkov radiation by particles moving faster than light in a medium. Note also similarity to the Landau criterion for superfluidity: In a fluid moving with the velocity $\mathbf{v}$, an obstacle or a wall can generate excitation with the momentum $\mathbf{p}$ and the energy $\varepsilon(p)$ if the resulting energy change is negative, $\varepsilon(p) + (\mathbf{p}\cdot\mathbf{v}) < 0$, which requires $v > \varepsilon(p)/p$. If the spectrum of excitations has a minimal $\varepsilon(p)/p$ then the fluid moving with $v < \min[\varepsilon(p)/p]$ meets no resistance. For gravity-capillary surface waves on deep water, the requirement for the Vavilov–Cherenkov resonance means that V must exceed the minimal phase velocity, which is

$$c(k_0) = \frac{\omega(k_0)}{k_0} = \left(\frac{4\alpha g}{\rho}\right)^{1/4} \simeq 23\,\mathrm{cm\,s}^{-1}. \tag{3.25}$$

When the phase velocity is minimal it coincides with the group velocity:

$$\frac{\partial}{\partial k}\frac{\omega}{k} = \frac{1}{k}\left(\frac{\partial \omega}{\partial k} - \frac{\omega}{k}\right) = 0.$$

Let us stress that the minimal group velocity v_* is smaller than $c(k_0)$ and corresponds to a larger wavelength, $\lambda_* > \lambda_0$, see Figure. Minimal phase velocity determines the threshold speed for generating waves. Minimal group velocity determines the speed of the caustic.

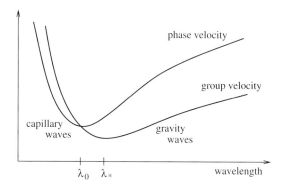

Consider first a source that is long in the direction perpendicular to the motion, say a tree fallen across the stream. Then a one-dimensional pattern of surface waves is generated with two wavenumbers that correspond to $c(k) = V$. These two wavenumbers correspond to the same phase velocity but different group velocities. The waves transfer energy from the source with the group velocity. Since the longer (gravity) wave has its group velocity lower, $\omega' < V$, then it is found behind the source (downstream), while the shorter (capillary) wave can be found upstream.[5] Of course, capillary waves can only be generated by a really thin object (much smaller than λ_*), like a fishing line or a small tree branch.

Generally, consider a medium where a wave with the wavenumber k has the frequency ω_k and a small decay rate γ_k. Assume that the wave source generating spectral density $A(k)$ started moving with the speed V relative to the medium at $x \to -\infty$ at time $t \to -\infty$ and is at $x = 0$ at $t = 0$. At that time the perturbation at every point x is the sum of the plane waves generated during the past:

$$\int_{-\infty}^{\infty} A(k)dk \int_{-\infty}^{0} dx' \int_{-\infty}^{0} dt' \delta(t' - x'/V)e^{ik(x-x')+i\omega_k t'+\gamma_k t'}$$

$$= \int_{-\infty}^{\infty} A(k)dk \int_{-\infty}^{0} dx' e^{ik(x-x')+(i\omega_k+\gamma_k)x'/V} = \int \frac{V i e^{ikx}dk}{kV - \omega_k + i\gamma_k} . \quad (3.26)$$

Since $\gamma_k \ll \omega(k)$, the main contribution into the integral is given by k close to $k_0 = \omega(k_0)/V$, which corresponds to the wave whose phase velocity coincides with the speed of the source. Then we can expand $\omega_k = \omega(k_0) + \omega'(k_0)\Delta k$ and write (3.26) as follows:

$$VA(k_0)e^{ik_0 x} \int_{-\infty}^{\infty} \frac{ie^{i\Delta kx}d\Delta k}{w\Delta k + i\gamma_0} . \tag{3.27}$$

Here $\gamma_0 = \gamma(k_0)$ and $w = V - \omega'(k_0) = \omega(k_0)/k_0 - \omega'(k_0)$ is the difference between the phase and group velocities. We also assumed that the source spectral density does not change when k changes by the values of the order γ/w, that is that the source size is less than w/γ. It is convenient to compute the integral (3.27) making the path a closed contour in the complex plane of Δk by adding a semicircle at infinity where the integrand is infinitesimal. For positive/negative x such semicircle is in the upper/lower half plane. The integrand has a pole at $\Delta k = -i\gamma_0/w$. When $w > 0$ the pole is in the lower half plane of complex Δk, so that the integral is zero for $x > 0$, while for $x < 0$ it is

$$\frac{2\pi V}{w} \exp(ik_0 x + \gamma_0 x/w) . \tag{3.28}$$

Indeed, when the phase velocity is larger than the group velocity the perturbation is behind the source. The same formula (3.28) describes the case $w < 0$ and $x > 0$. The temporal decay rate γ_k determines how the perturbation decays away from the source in space.

Ships generate an interesting pattern of gravity waves on the water surface that can be understood as follows. The wave generated at the angle θ to the ship's motion has its wavelength determined by the condition $V\cos\theta = c(k)$, necessary for the ship's bow to stay on the wave crest. This condition means that different wavelengths are generated at different angles in the interval $0 \leq \theta \leq \pi/2$. Similar to our consideration in Section 2.3.5, let us find the locus of points reached by the waves generated at A during the time it takes for the ship to reach B, see Figure 3.3. The waves propagate away from the source with group velocity $\omega' = c(k)/2$. The fastest is the wave generated in the direction of the ship propagation ($\theta = 0$) which moves with group velocity equal to half the ship's speed and reaches A' such that AA' = AB/2. The wave generated at the angle θ reaches E such that $AE = AA'\cos\theta$ which means that AEA' is a right angle. We conclude that the waves generated at different angles reach the circle with the diameter AA'. Since OB = 3OC, all the waves generated before the ship reached B are within the *Kelvin wedge* with the angle $\varphi = \arcsin(1/3) \approx 19.5°$; compare this with the Mach cone shown in Figure 2.14. Note the remarkable fact that the angle of the Kelvin wedge is completely universal, i.e. independent of the ship's speed.[6]

Let us describe now the form of the wave crests, which are neither straight nor parallel since they are produced by the waves emitted at different moments

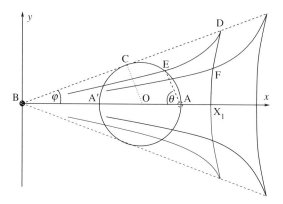

Figure 3.3 Pattern of waves generated by a ship that moves along the x-axis to the left. The circle is the locus of points reached by waves generated at A when the ship arrives at B. The broken lines show the Kelvin wedge. The solid lines are wave crests.

at different angles. The wave propagating at the angle θ makes a crest at the angle $\pi/2 - \theta$. Consider point E on a crest with the coordinates

$$x = AB(2 - \cos^2\theta)/2, \quad y = (AB/2)\sin\theta\cos\theta.$$

The crest slope must satisfy the equation $dy/dx = \cot\theta$, which gives $dAB/d\theta = -AB\tan\theta$. The solution of this equation, $AB = X_1\cos\theta$, describes how the source point A is related to the angle θ of the wave that creates the given crest. Different integration constants X_1 correspond to different crests. The crest shape is given parametrically by

$$x = X_1\cos\theta(2 - \cos^2\theta)/2, \quad y = (X_1/2)\cos\theta\sin 2\theta \qquad (3.29)$$

and it is shown in Figure 3.3 by the solid lines, see also Figure 3.4. As expected, longer (faster) waves propagate at smaller angles. Note that for every point inside the Kelvin wedge one can find two different source points. That means that two constant-phase lines cross at every point, like the crossing of crests seen at the point F. One sees two distinct families of waves: diverging from the ship and transverse to the direction of motion. There are no waves outside the Kelvin wedge, whose boundary is thus a caustic,[7] where every crest has a cusp like at the point D in Figure 3.3, determined by the condition that both $x(\theta)$ and $y(\theta)$ have maxima. Differentiating (3.29) and solving $dx/d\theta = dy/d\theta = 0$, we obtain the propagation angle $\cos^2\theta_0 = 2/3$, which is actually the angle CAB since it corresponds to the wave that reached the wedge. We can express the angle COA alternatively as $\pi/2 + \varphi = \pi - 2\theta_0$ and relate: $\theta_0 = \pi/4 - \varphi/2 \approx 35°$.

Figure 3.4 The wave pattern of a ship consists of the breaking wave from the bow with its turbulent wake (distinguished by a short trace of white foam) and the Kelvin pattern. Inside the Kelvin wedge one can see two lines of maxima created by wavelengthes comparable to the ship length. Photograph copyright: Alexey Baskakov, www.dreamstime.com.

3.2 Weakly nonlinear waves

The law of linear wave propagation is completely characterized by the dispersion relation ω_k. It does not matter what physical quantity oscillates in the wave (fluid velocity, density, electromagnetic field, surface elevation, etc.), waves with the same dispersion relation propagate in the same way. Can we achieve the same level of universality in describing nonlinear waves? As we shall see in this and the next sections, some universality classes can be distinguished but the level of universality naturally decreases as nonlinearity increases.

3.2.1 Hamiltonian description

What else, apart from ω_k, must we know to describe weakly nonlinear waves? Since every wave is determined by two dynamic variables, amplitude and phase, it is natural to employ the Hamiltonian formalism where variables also come in pairs (coordinate–momentum, action–angle). That must work for waves in a conservative medium. Indeed, the Hamiltonian formalism is the most general way to describe systems that satisfy the least-action principle, as most closed physical systems do. The main advantage of the Hamiltonian formalism (compared, say, with its particular case, the Lagrangian formalism) is an ability to use canonical transformations. Those transformations involve both coordinates and momenta and are thus more general than the coordinate transformations one uses within the Lagrangian formalism, which employs coordinates and their time derivatives (do not confuse Lagrangian formalism in mechanics and field theory with the Lagrangian description in fluid mechanics). Canonical transformation is a powerful tool that allows one to reduce a variety of problems into a few universal problems. So let us try to understand what is a general form of the Hamiltonian of a weakly nonlinear wave system.

As we have seen in Section 1.3.4, Hamiltonian mechanics of continuous systems lives in an even-dimensional space of coordinates $q(\mathbf{r},t)$ and momenta $\pi(\mathbf{r},t)$ that satisfy the equations

$$\frac{\partial q(\mathbf{r},t)}{\partial t} = \frac{\delta \mathcal{H}}{\delta \pi(\mathbf{r},t)}, \qquad \frac{\partial \pi(\mathbf{r},t)}{\partial t} = -\frac{\delta \mathcal{H}}{\delta q(\mathbf{r},t)}.$$

Here the Hamiltonian $\mathcal{H}\{q(\mathbf{r},t),\pi(\mathbf{r},t)\}$ is a functional (simply speaking, a function presents a number for every number, while a functional presents a number for every function; for example, a definite integral is a functional). The variational derivative $\delta/\delta f(r)$ is a generalization of the partial derivative $\partial/\partial f(r_n)$ from a discrete to a continuous set of variables. The variational derivative of a linear functional of the form $I\{f\} = \int \phi(r')f(r')\,dr'$ is calculated by

$$\frac{\delta I}{\delta f(r)} = \int \phi(r') \frac{\delta f(r')}{\delta f(r)}\, dr' = \int \phi(r')\delta(r-r')\, dr' = \phi(r).$$

For this, one mentally replaces $\delta/\delta f(r)$ with $\partial/\partial f(r_n)$ and integration with summation.

For example, the Euler and continuity equations for potential flows (in particular, acoustic waves) can be written in a Hamiltonian form:

$$\frac{\partial \rho}{\partial t} = \frac{\delta \mathcal{H}}{\delta \phi}, \quad \frac{\partial \phi}{\partial t} = -\frac{\delta \mathcal{H}}{\delta \rho}, \tag{3.30}$$

$$\mathcal{H} = \int \rho \left[\frac{|\nabla \phi|^2}{2} + E(\rho) \right] d\mathbf{r}.$$

We shall use canonical variables even more symmetrical than π, q, analogous to what are called creation–annihilation operators in quantum theory. In our case, they are just functions, not operators. Assuming p and q to be of the same dimensionality (which can always be achieved by multiplying them by factors) we introduce

$$a = (q + i\pi)/\sqrt{2}, \qquad a^* = (q - i\pi)/\sqrt{2}.$$

Instead of two real equations for p, q we now have one complex equation

$$i\frac{\partial a(\mathbf{r}, t)}{\partial t} = \frac{\delta \mathcal{H}}{\delta a^*(\mathbf{r}, t)}. \tag{3.31}$$

The complex conjugated equation describes the evolution of a^*.

In the linear approximation, waves with different wavevectors do not interact and their equations are independent. Normal canonical coordinates in infinite space are the complex Fourier amplitudes a_k, which satisfy the equation

$$\frac{\partial a_k}{\partial t} = -i\omega_k a_k.$$

Comparing this with (3.31) we conclude that the Hamiltonian of a linear wave system is quadratic in the amplitudes:

$$\mathcal{H}_2 = \int \omega_k |a_k|^2 \, d\mathbf{k}. \tag{3.32}$$

It is the energy density per unit volume. Terms of higher orders describe wave interaction due to nonlinearity; the lowest terms are cubic:

$$\mathcal{H}_3 = \int \left[\left(V_{123} a_1^* a_2 a_3 + \text{c.c.} \right) \delta(\mathbf{k}_1 - \mathbf{k}_2 - \mathbf{k}_3) \right.$$
$$\left. + \left(U_{123} a_1^* a_2^* a_3^* + \text{c.c.} \right) \delta(\mathbf{k}_1 + \mathbf{k}_2 + \mathbf{k}_3) \right] d\mathbf{k}_1 d\mathbf{k}_2 d\mathbf{k}_3. \tag{3.33}$$

Here c.c. means the complex conjugated terms and we use shorthand notations $a_1 = a(\mathbf{k}_1)$, etc. The delta functions express momentum conservation and appear because of space homogeneity. Indeed, (3.32) and (3.33) are, respectively, the Fourier representation of the integrals, like

$$\int \Omega(\mathbf{r}_1 - \mathbf{r}_2) a(\mathbf{r}_1) a(\mathbf{r}_2) \, d\mathbf{r}_1 d\mathbf{r}_2$$

and

$$\int V(\mathbf{r}_1 - \mathbf{r}_2, \mathbf{r}_1 - \mathbf{r}_3) a(\mathbf{r}_1) a(\mathbf{r}_2) a(\mathbf{r}_3) \, d\mathbf{r}_1 d\mathbf{r}_2 d\mathbf{r}_3.$$

The Hamiltonian is real and its coefficients have obvious symmetries, $U_{123} = U_{132} = U_{213}$ and $V_{123} = V_{132}$. We presume that every next term in the Hamiltonian expansion is smaller than the previous one, in particular, $\mathcal{H}_2 \gg \mathcal{H}_3$, which requires

$$\omega_k \gg V |a_k| k^d, \quad U |a_k| k^d, \tag{3.34}$$

where d is space dimensionality (two for surface waves and three for sound, for instance).

In the perverse nature of people who learnt quantum mechanics before fluid mechanics, we may use an analogy between a, a^* and the quantum creation–annihilation operators a, a^+ and suggest that the V-term must describe the confluence $2 + 3 \rightarrow 1$ and the reverse process of decay $1 \rightarrow 2 + 3$. Similarly, the U-term must describe the creation of three waves from a vacuum and the opposite process of annihilation. We shall use quantum analogies quite often in this chapter since quantum physics is to a large extent a wave physics. To support this quantum-mechanical interpretation and make explicit the physical meaning of different terms in the Hamiltonian, we write a general equation of motion for weakly nonlinear waves:

$$\frac{\partial a_k}{\partial t} = -3\mathrm{i} \int U_{k12} a_1^* a_2^* \delta(\mathbf{k}_1 + \mathbf{k}_2 + \mathbf{k}) \, \mathrm{d}\mathbf{k}_1 \mathrm{d}\mathbf{k}_2 \qquad (3.35)$$

$$- \mathrm{i} \int V_{k12} a_1 a_2 \delta(\mathbf{k}_1 + \mathbf{k}_2 - \mathbf{k}) \, \mathrm{d}\mathbf{k}_1 \mathrm{d}\mathbf{k}_2$$

$$- 2\mathrm{i} \int V_{1k2}^* a_1 a_2^* \delta(\mathbf{k}_1 - \mathbf{k}_2 - \mathbf{k}) \, \mathrm{d}\mathbf{k}_1 \mathrm{d}\mathbf{k}_2 - (\gamma_k + \mathrm{i}\omega_k) a_k,$$

where we have also included the linear damping γ_k (as one always does when resonances are possible). The delta functions in the integrals suggest that each respective term describes the interaction between three different waves. To see this explicitly, consider a particular initial condition with two waves, having, respectively, wavevectors $\mathbf{k}_1, \mathbf{k}_2$, frequencies ω_1, ω_2 and finite amplitudes A_1, A_2. Then the last nonlinear term in (3.35), $-2\mathrm{i} e^{\mathrm{i}(\omega_2 - \omega_1)t} V_{1k2}^* A_1 A_2^* \delta(\mathbf{k}_1 - \mathbf{k}_2 - \mathbf{k})$, provides the wave $\mathbf{k} = \mathbf{k}_1 - \mathbf{k}_2$ with periodic forcing of frequency $\omega_1 - \omega_2$, and similarly the other two terms. The forced solution of (3.35) is then as follows:

$$a(\mathbf{k}, t) = -3\mathrm{i} e^{\mathrm{i}(\omega_1 + \omega_2)t} \frac{U_{k12} A_1^* A_2^* \delta(\mathbf{k}_1 + \mathbf{k}_2 + \mathbf{k})}{\gamma_k + \mathrm{i}(\omega_1 + \omega_2 + \omega_k)}$$

$$- \mathrm{i} e^{-\mathrm{i}(\omega_1 + \omega_2)t} \frac{V_{k12} A_1 A_2 \delta(\mathbf{k}_1 + \mathbf{k}_2 - \mathbf{k})}{\gamma_k + \mathrm{i}(\omega_1 + \omega_2 - \omega_k)}$$

$$- 2\mathrm{i} e^{\mathrm{i}(\omega_2 - \omega_1)t} \frac{V_{1k2}^* A_1 A_2^* \delta(\mathbf{k} + \mathbf{k}_2 - \mathbf{k}_1)}{\gamma_k + \mathrm{i}(\omega_k + \omega_2 - \omega_1)}.$$

Here and below, $\omega_{1,2} = \omega(\mathbf{k}_{1,2})$. Because of (3.34) the amplitudes of the secondary waves are small except for the cases of resonances, that is when the driving frequency coincides with the eigenfrequency of the wave with the respective k. The amplitude $a(\mathbf{k}_1 + \mathbf{k}_2)$ is not small if $\omega(\mathbf{k}_1 + \mathbf{k}_2) + \omega(\mathbf{k}_1) + \omega(\mathbf{k}_2) = 0$. This can happen in the nonequilibrium medium where negative-frequency waves are possible. Negative frequency corresponds to negative energy (3.32), which means that excitation of the wave decreases the energy of the medium. This may be the case, for instance, when there are currents in the medium and the wave moves against the current. In nonequilibrium medium, the frequency can also be complex (which signals instability), then $\mathcal{H}_2$ is different from (3.32); see Exercise 3.4. Two other resonances require $\omega(\mathbf{k}_1 + \mathbf{k}_2) = \omega(\mathbf{k}_1) + \omega(\mathbf{k}_2)$ and $\omega(\mathbf{k}_1 - \mathbf{k}_2) = \omega(\mathbf{k}_1) - \omega(\mathbf{k}_2)$; the dispersion relations that allow for this are called dispersion relations of the decay type. For example, the power dispersion relation $\omega_k \propto k^\alpha$ is of the decay type if $\alpha \geq 1$ and of the nondecay type (that is, does not allow for the three-wave resonance) if $\alpha < 1$, see Exercise 3.5.

3.2.2 Hamiltonian normal forms

Intuitively, it is clear that nonresonant processes are unimportant for weak non-linearity. Technically, one can use the canonical transformations to eliminate the nonresonant terms from the Hamiltonian. Because the terms that we want to eliminate are small, the transformation should be close to identical. Consider some continuous distribution of a_k. If one wants to get rid of the U-term in $\mathcal{H}_3$, one makes the following transformation

$$b_k = a_k - 3 \int \frac{U_{k12} a_1^* a_2^*}{\omega_1 + \omega_2 + \omega_k} \delta(\mathbf{k}_1 + \mathbf{k}_2 + \mathbf{k}) \, d\mathbf{k}_1 d\mathbf{k}_2. \qquad (3.36)$$

It is possible when the denominator does not turn into zero in the integration domain determined by the delta-function of the wave vectors, that is when the spatia-temporal resonance is impossible. This is the case, in particular, for all media that were in thermal equilibrium before wave excitation. The Hamiltonian $\mathcal{H}\{b, b^*\} = \mathcal{H}_2 + \mathcal{H}_3$ does not contain the U-term. The elimination of the V-terms is made by a similar transformation

$$b_k = a_k + \int \left[\frac{V_{k12} a_1 a_2 \delta(\mathbf{k}_1 + \mathbf{k}_2 - \mathbf{k})}{\omega_1 + \omega_2 - \omega_k} + \frac{V_{1k2}^* a_1 a_2^* \delta(\mathbf{k}_1 - \mathbf{k}_2 - \mathbf{k})}{\omega_1 - \omega_2 - \omega_k} \right.$$

$$\left. + \frac{V_{2k1}^* a_1^* a_2 \delta(\mathbf{k}_2 - \mathbf{k}_1 - \mathbf{k})}{\omega_2 - \omega_1 - \omega_k} \right] d\mathbf{k}_1 d\mathbf{k}_2, \qquad (3.37)$$

possible only for the nondecay dispersion relation. We see that both transformations (3.36, 3.37) are possible when denominators do not turn into zero, that is when the respective processes are nonresonant. One can check that the transformations are canonical, that is $i\dot{b}_k = \delta \mathcal{H}\{b, b^*\}/\delta b_k^*$. The procedure described here was invented for excluding nonresonant terms in celestial mechanics and later generalized for continuous systems.[8]

We may thus conclude that (3.33) is the proper Hamiltonian of interaction only when all three-wave processes are resonant. When there are no negative-energy waves but the dispersion relation is of the decay type (like for capillary waves on deep water), the proper interaction Hamiltonian contains only the V-term. When the dispersion relation is of the nondecay type (like for gravity waves on water), all the cubic terms can be excluded and the interaction Hamiltonian must be of the fourth order in wave amplitudes. Moreover, if the dispersion relation does not allow $\omega(\mathbf{k}_1 + \mathbf{k}_2) = \omega(\mathbf{k}_1) + \omega(\mathbf{k}_2)$ then it does not allow $\omega(\mathbf{k}_1 + \mathbf{k}_2 + \mathbf{k}_3) = \omega(\mathbf{k}_1) + \omega(\mathbf{k}_2) + \omega(\mathbf{k}_3)$ as well. This means that when decays of the type $1 \to 2 + 3$ are nonresonant, four-wave decays like $1 \to 2 + 3 + 4$ are nonresonant too. So the proper Hamiltonian in this case describes four-wave scattering $1 + 2 \to 3 + 4$, which is always resonant:

$$\mathcal{H}_4 = \int T_{1234} a_1 a_2 a_3^* a_4^* \delta(\mathbf{k}_1 + \mathbf{k}_2 - \mathbf{k}_3 - \mathbf{k}_4)\, d\mathbf{k}_1 d\mathbf{k}_2 d\mathbf{k}_3 d\mathbf{k}_4 \,. \qquad (3.38)$$

One may ask: why bother with transformations (3.36,3.37) and not just omit nonresonant terms from the hamiltonian? The point is that in the new variables the remaining interaction coefficients change. For example, after excluding cubic terms, T_{1212} acquires additions of the type $|V_{1+2,12}|^2/(\omega_{1+2} - \omega_1 - \omega_2)$. If there are surfaces in the space $\{\mathbf{k}_1, \mathbf{k}_2\}$ where the denominator $\omega_{1+2} - \omega_2$ is small (i.e. the cubic processes are almost resonant, as for sound waves with small positive dispersion), such additions could be dominant.

3.2.3 Wave instabilities

Wave motion itself can be unstable with respect to small perturbations. Let us show that if the dispersion relation is of the decay type, then a monochromatic wave of sufficiently high amplitude is subject to a *decay instability*. Consider the medium that contains a finite-amplitude wave $A \exp[i(\mathbf{kr} - \omega_k t)]$ and add initial perturbations in the form of two waves with small amplitudes a_1, a_2. For interaction to have a net effect, all three waves must have wavevectors in a spatial resonance: $\mathbf{k}_1 + \mathbf{k}_2 = \mathbf{k}$. We write linearized equations on perturbation leaving only resonant terms in (3.35):

$$\dot{a}_1 + (\gamma_1 + i\omega_1)a_1 + 2iV_{k12}A a_2^* \exp(-i\omega_k t) = 0\,,$$
$$\dot{a}_2^* + (\gamma_2 - i\omega_2)a_2^* + 2iV_{k12}^* A^* a_1 \exp(i\omega_k t) = 0\,.$$

The solution can be sought in the form

$$a_1(t) \propto \exp(\Gamma t - i\Omega_1 t)\,, \quad a_2^*(t) \propto \exp(\Gamma t + i\Omega_2 t)\,.$$

The temporal resonance condition is $\Omega_1 + \Omega_2 = \omega_k$. The amplitudes of the waves will be determined by the mismatches, $\Omega_1 - \omega_1$ and $\Omega_2 - \omega_2$, which are the differences between the forced frequencies and the eigenfrequencies, determined by dispersion relation. The sum of the mismatches is $\Delta\omega = \omega_1 + \omega_2 - \omega_k$. We are interested in Ωs that give maximal real Γ and expect it when 1 and 2 are symmetrical, so that $\Omega_1 - \omega_1 = \Omega_2 - \omega_2 = \Delta\omega/2$. Consider, for simplicity, $\gamma_1 = \gamma_2$, then

$$\Gamma = -\gamma \pm \sqrt{4|V_{k12}A|^2 - (\Delta\omega)^2/4}\,. \tag{3.39}$$

If the dispersion relation is of the nondecay type, then $\Delta\omega \simeq \omega_k \gg |VA|$ and there is no instability. On the contrary, for decay dispersion relations, resonance is possible, $\Delta\omega = \omega_1 + \omega_2 - \omega_k = 0$, so that the growth rate of instability, $\Gamma = 2|V_{k12}A| - \gamma$, is positive when the amplitude is larger than the threshold: $A > \gamma/2|V_{k12}|$, i.e. when the nonlinearity overcomes dissipation. The growth rate is maximal for those a_1, a_2 that are in resonance (i.e. $\Delta\omega = 0$) and have minimal $(\gamma_1 + \gamma_2)/|V_{k12}|$. In the particular case $k = 0$, $\omega_0 \neq 0$, decay instability is called parametric instability, since it corresponds to a periodic uniform change in some system parameter. For example, Faraday discovered that a vertical vibration of a container with a fluid leads to the parametric excitation of a standing surface wave ($k_1 = -k_2$) with half the vibration frequency (the parameter being changed periodically is the gravity acceleration g). In a simple case of an oscillator it is called parametric resonance, known to any child on a swing who stretches and folds his legs with twice the frequency of the swing (the parameter being changed periodically is the swing effective length L). In both cases, one varies the frequency $\sqrt{g/L}$, which is the parameter of the Hamiltonian.

As with any instability, the usual question is what stops the exponential growth and the usual answer is that further nonlinearity does that, as in Section 2.1.3. When the amplitude is not far from the threshold, these nonlinear effects can be described in the mean-field approximation as the renormalization of the linear parameters ω_k, γ_k and of the pumping $V_{k12}A$. The renormalization should be such as to put the wave system back to the threshold, that is to turn the renormalized Γ into zero. The frequency renormalization $\tilde{\omega}_k = \omega_k + \int T_{kk'kk'}|a_{k'}|^2 \, \mathrm{d}k'$ (see the next section) appears because of the four-wave processes and can take waves out of resonance if the set of wavevectors is discrete, owing to a finite box size (it is a mechanism of instability restriction for finite-dimensional systems like an oscillator – swing frequency decreases with amplitude, for instance). If, however, the box is large enough, then the frequency spectrum is close to continuous and there are waves in resonance for any nonlinearity. In this case, the saturation of instability is caused by renormalization of the damping and pumping. The renormalization (increase) of γ_k appears because of the waves of the third generation that take energy from a_1, a_2. The pumping renormalization appears because of the four-wave interaction, for example, (3.38) adds $-ia_2^* \int T_{1234} a_3 a_4 \delta(\mathbf{k} - \mathbf{k}_3 - \mathbf{k}_4) \, d\mathbf{k}_3 d\mathbf{k}_4$ to $\dot{a}_1$.

3.3 Nonlinear Schrödinger equation (NSE)

This section is devoted to a nonlinear spectrally narrow wave packet. Consideration of the linear propagation of such a packet in Section 3.1.4 taught us the notions of phase and group velocities and caustics. In this section, the account of nonlinearity brings equally fundamental notions of the Bogoliubov spectrum of condensate fluctuations, modulational instability, solitons, self-focusing, collapse and wave turbulence.

3.3.1 Derivation of NSE

Consider a quasimonochromatic wave packet in an isotropic nonlinear medium. Since the wave amplitudes are nonzero in a narrow region Δk of $\mathbf{k}$-space around some $\mathbf{k}_0$, then the processes changing the number of waves (like $1 \to 2 + 3$ and $1 \to 2 + 3 + 4$) are nonresonant because the frequencies of all waves are close. Therefore, all the nonlinear terms can be eliminated from the interaction Hamiltonian except $\mathscr{H}_4$ and the equation of motion has the form

$$\frac{\partial a_k}{\partial t} + i\omega_k a_k = -i \int T_{k123} a_1^* a_2 a_3 \delta(\mathbf{k} + \mathbf{k}_1 - \mathbf{k}_2 - \mathbf{k}_3) \, d\mathbf{k}_1 d\mathbf{k}_2 d\mathbf{k}_3. \quad (3.40)$$

Consider now $\mathbf{k} = \mathbf{k}_0 + \mathbf{q}$ with $q \ll k_0$ and expand, similar to (3.18),

$$\omega(k) = \omega_0 + (\mathbf{q}\mathbf{v}) + \frac{1}{2} q_i q_j \left(\frac{\partial^2 \omega}{\partial k_i \partial k_j} \right)_0,$$

where $\mathbf{v} = \partial \omega / \partial \mathbf{k}$ at $k = k_0$. In an isotropic medium ω depends only on modulus k and

$$q_i q_j \frac{\partial^2 \omega}{\partial k_i \partial k_j} = q_i q_j \frac{\partial}{\partial k_i} \frac{k_j}{k} \frac{\partial \omega}{\partial k} = q_i q_j \left[\frac{k_i k_j \omega''}{k^2} + \left(\delta_{ij} - \frac{k_i k_j}{k^2} \right) \frac{v}{k} \right]$$

$$= q_\parallel^2 \omega'' + \frac{q_\perp^2 v}{k}.$$

Let us introduce the temporal envelope $a_k(t) = \exp(-i\omega_0 t) \psi(\mathbf{q}, t)$ into (3.40):

$$\left[i \frac{\partial}{\partial t} - (\mathbf{q}\mathbf{v}) - \frac{q_\parallel^2 \omega''}{2} - \frac{q_\perp^2 v}{2k} \right] \psi_q = T \int \psi_1^* \psi_2 \psi_3 \delta(\mathbf{q} + \mathbf{q}_1 - \mathbf{q}_2 - \mathbf{q}_3) \, d\mathbf{q}_1 d\mathbf{q}_2 d\mathbf{q}_3.$$

We assumed the nonlinear term to be small, $T|a_k|^2 (\Delta k)^{2d} \ll \omega_k$, and took it at $k = k_0$. This result is usually represented in r-space for $\psi(\mathbf{r}) = \int \psi_q \exp(i\mathbf{q}\mathbf{r}) \, d\mathbf{q}$. The nonlinear term is local in r-space:

$$
\int d\mathbf{r}_1 d\mathbf{r}_2 d\mathbf{r}_3 \, \psi^*(\mathbf{r}_1) \psi(\mathbf{r}_2) \psi(\mathbf{r}_3) \int d\mathbf{q} d\mathbf{q}_1 d\mathbf{q}_2 d\mathbf{q}_3 \delta(\mathbf{q} + \mathbf{q}_1 - \mathbf{q}_2 - \mathbf{q}_3)
$$

$$
\times \exp\left[i(\mathbf{q}_1 \mathbf{r}_1) - i(\mathbf{q}_2 \mathbf{r}_2) - i(\mathbf{q}_3 \mathbf{r}_3) + i(\mathbf{q}\mathbf{r})\right]
$$

$$
= \int d\mathbf{r}_1 d\mathbf{r}_2 d\mathbf{r}_3 \, \psi^*(\mathbf{r}_1) \psi(\mathbf{r}_2) \psi(\mathbf{r}_3) \delta(\mathbf{r}_1 - \mathbf{r}) \delta(\mathbf{r}_2 - \mathbf{r}) \delta(\mathbf{r}_3 - \mathbf{r}) = |\psi|^2 \psi,
$$

and the equation takes the form

$$
\frac{\partial \psi}{\partial t} + v \frac{\partial \psi}{\partial z} - \frac{i\omega''}{2} \frac{\partial^2 \psi}{\partial z^2} - \frac{iv}{2k} \Delta_\perp \psi = -iT|\psi|^2 \psi.
$$

Here the term $v\partial_z$ is responsible for propagation with the group velocity, $\omega'' \partial_{zz}$ for dispersion and $(v/k)\Delta_\perp$ for diffraction. One may ask why in the expansion of ω_{k+q} we kept the terms both linear and quadratic in small q. This is because the linear term (which gives $\partial \psi / \partial z$ in the last equation) can be eliminated by the transition to the moving reference frame $z \to z - vt$. We also renormalize the transversal coordinate by the factor $\sqrt{k_0 \omega''/v}$ and obtain the celebrated nonlinear Schrödinger equation

$$
i\frac{\partial \psi}{\partial t} + \frac{\omega''}{2} \Delta \psi - T|\psi|^2 \psi = 0. \tag{3.41}
$$

Sometimes (particularly for $T < 0$) it is called the Gross–Pitaevsky equation after the scientists who derived it to describe a quantum condensate. This equation is meaningful at different dimensionalities. It may describe the evolution of a three-dimensional packet, as in a Bose–Einstein condensation of cold atoms. When $\mathbf{r}$ is two-dimensional, it may correspond either to the evolution of the packet in a 2D medium (say, for surface waves) or to steady propagation in 3D described by $iv\psi_z + (v/2k)\Delta_\perp \psi = T|\psi|^2 \psi$, which turns into (3.41) upon re-labeling $z \to vt$. In a steady case, one neglects ψ_{zz} since this term is much less than ψ_z. In a nonsteady case, this is not necessarily so, since ∂_t and $v\partial_z$ might be about to annihilate each other and one is interested in the next terms. And, finishing with dimensionalities, the one-dimensional NSE corresponds to a stationary two-dimensional case or evolution in a one-dimensional medium.

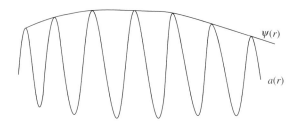

Different media provide for different signs of the coefficients. Apart from hydrodynamic applications, the NSE also describes nonlinear optics. Indeed, Maxwell's equation for waves takes the form $[\omega^2 - (c^2/n)\Delta]E = 0$. The refraction index depends on the wave intensity: $n = 1 + 2\alpha|E|^2$. There are different reasons for that dependence (and so different signs of α may be realized in different materials), for example: electrostriction, heating and the Kerr effect (orientation of nonisotropic molecules by the wave field). We consider waves moving mainly in one direction and pass into the reference frame moving with the velocity c, i.e. change $\omega \to \omega - ck$. Expanding in small parameters $k_\perp/k$ and $\alpha|E|^2$,

$$ck/\sqrt{n} \approx ck(1 - \alpha|E|^2) + ck_\perp^2/2k,$$

substituting it into

$$(\omega - ck - ck/\sqrt{n})(\omega - ck + ck/\sqrt{n})E = 0,$$

and retaining only the first nonvanishing terms in diffraction and nonlinearity, we obtain the NSE with $v = c$ after the inverse Fourier transform $\omega \to i\partial_t$, $k_\perp^2 \to -\Delta_\perp$. In particular, the one-dimensional NSE describes light in an optical fibre, sound in a beam and pulse in a nerve.

3.3.2 Modulational instability

The simplest effect of the four-wave scattering is frequency renormalization. Indeed, the NSE has a stationary solution as a plane wave with a renormalized frequency $\psi_0(t) = A_0 \exp(-iTA_0^2 t)$ (in quantum physics, this state, coherent across the whole system, corresponds to a Bose–Einstein condensate). Let us describe small perturbations of the condensate. We write the perturbed solution as $\psi = (A_0 + \tilde{A})e^{-iTA_0^2 t + i\varphi}$ and assume the perturbation to be one-dimensional (along the direction which we denote ξ). The real and imaginary parts of the linearized NSE take the form

$$\tilde{A}_t + \frac{\omega''}{2}A_0\varphi_{\xi\xi} = 0, \quad \varphi_t = -2TA_0\tilde{A} + \frac{\omega''}{2A_0}\tilde{A}_{\xi\xi}.$$

If the amplitude of the perturbation is modulated then so is the phase:

$$\tilde{A} = \propto \exp(\imath k\xi - \imath\Omega t), \quad \varphi \propto \exp(\imath k\xi - \imath\Omega t).$$

The dispersion relation for the perturbations then takes the form:

$$\Omega^2 = T\omega'' A_0^2 k^2 + \omega''^2 k^4/4. \tag{3.42}$$

When $T\omega'' > 0$, it is called the Bogoliubov formula for the spectrum of condensate perturbations. We have an instability when $T\omega'' < 0$ (the Lighthill criterion). I first explain this criterion using the language of classical waves and at the end of the section give an alternative explanation in terms of quantum (quasi)particles. Classically, we define the frequency as minus the time derivative of the phase: $\varphi_t = -\omega$. For a nonlinear wave, the frequency is generally dependent on both the amplitude and the wavenumber. The first derivatives of the frequency with respect to the amplitude and the wavenumber are zero at zero for systems with the symmetry $k \to -k$ and $A \to -A$. The factors T and ω'' are the second derivatives, respectively, with respect to the amplitude and the wavenumber. That is, instability happens when the surface $\omega(k, A)$ has a saddle point at $k = 0 = A$. From the spectral viewpoint, we look at a four-wave interaction between the carrier wave and the perturbations called side bands. For that interaction to be resonant, the dispersive correction to the side-band frequency needs to be compensated by the nonlinearity correction. From a spatiatemporal viewpoint, let us describe the evolution of a local amplitude minimum when, for instance, $\omega'' > 0$ and $T < 0$. The frequency has a maximum in the amplitude minimum. Consider the current wavenumber $K = \varphi_\xi$, whose time derivative is as follows: $K_t = \varphi_{\xi t} = -\omega_\xi$. The local maximum in ω means that K_t changes sign, that is K will grow to the right of the ω maximum and decrease to the left of it. The group velocity ω' grows with K since $\omega'' > 0$. Then the group velocity grows to the right and decreases to the left so that the parts separate (as the arrows show) and the perturbation deepens, as shown in Figure 3.5.

The result of this instability can be seen on the beach, where waves coming to the shore are modulated. Indeed, for long water waves $\omega_k \propto \sqrt{k}$ so that $\omega'' < 0$. As opposed to a pendulum and somewhat counterintuitively, the frequency grows with the amplitude and $T > 0$; it is related to the change of wave shape from sinusoidal to that forming a sharpened crest, which reaches $120°$ for sufficiently high amplitudes. A long water wave is thus unstable with respect to longitudinal modulations (Benjamin–Feir instability 1967). The growth rate is maximal for $k = A_0\sqrt{-2T/\omega''}$, which depends on the amplitude (Figure 3.6). Still, folklore has it that approximately every ninth wave is the largest. The maximal growth rate is quadratic in the wave amplitude: $\mathrm{Im}\,\Omega = TA_0^2$.

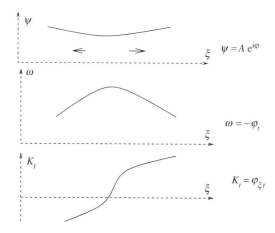

Figure 3.5 Space dependencies of the wave amplitude, frequency and time derivative of the wavenumber, which demonstrate the mechanism of the modulational instability for $\omega'' > 0$ and $T < 0$.

For transverse propagation of perturbations, one has to replace ω'' by v/k, which is generally positive so the criterion of instability is $T < 0$ or $\partial \omega / \partial |a|^2 < 0$, which also means that for instability the wave velocity has to decrease with amplitude. This can be easily visualized: if the wave is transversely modulated then the parts of the front where the amplitude is larger will move slower and further increase the amplitude because of focusing from neighboring parts, as shown in Figure 3.7.

Let us now find a quantum explanation for the modulational instability. Remember that the NSE (3.41) is a Hamiltonian system ($i\psi_t = \delta \mathcal{H}/\delta \psi^*$) with

$$\mathcal{H} = \frac{1}{2} \int \left(\omega'' |\nabla \psi|^2 + T |\psi|^4 \right) d\mathbf{r}. \qquad (3.43)$$

The Lighthill criterion means that the modulational instability happens when the Hamiltonian is not sign-definite. The overall sign of the Hamiltonian is unimportant, as the Hamiltonian dynamics are time-reversible and one can always change $\mathcal{H} \to -\mathcal{H}$, $t \to -t$; it is important that different configurations of $\psi(\mathbf{r})$ give different signs of the Hamiltonian because its two terms have different signs. As a result, the uniform state is unstable with respect to breaking into regions where one of the terms dominate. Consider $\omega'' > 0$. Using the quantum language one can interpret the first term in the Hamiltonian as the kinetic energy of (quasi)particles and the second term as their potential energy. For $T < 0$, the interaction is attractive, which leads to the instability. For the condensate, the kinetic energy (or pressure) is balanced by the interaction; a

Figure 3.6 Disintegration of the periodic wave due to modulational instability as demonstrated experimentally by Benjamin and Feir (1967). The upper photograph shows a regular wave pattern close to a wavemaker. The lower photograph is made some 60 meters (28 wavelengths) away, where the wave amplitude is comparable, but spatial periodicity is lost. The instability was triggered by imposing on the periodic motion of the wavemaker a slight modulation at the unstable side-band frequency; the same disintegration occurs naturally over longer distances. Photograph by J. E. Feir, reproduced from *Proc. R. Soc. Lond. A*, **299**, 59 (1967).

local perturbation with more particles (higher $|\psi|^2$) will make the interaction stronger, which leads to the contraction of perturbation and further growth of $|\psi|^2$.

Let us compare the modulational instability due to four-wave interaction with the decay instability due to three-wave interaction. The former presumes Lighthill criterium while the latter takes place for any interaction coefficient V_{k12} not turning into zero at the resonant manifold $\omega(\mathbf{k}_1 + \mathbf{k}_2) = \omega(k_1) + \omega(k_2)$. The growth rate of the decay instability $V_{k12}A$ is linear in the wave amplitude,

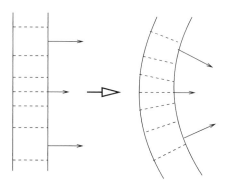

Figure 3.7 Transverse instability for the velocity decreasing with the amplitude.

while the modulational instability appears in the next order, so its growth rate TA^2 is generally smaller. For waves on deep water, decay instability is possible for short (capillary) waves; decays are getting nonresonant starting from some wavelength, so that longer (gravity) waves are subject to modulational instability.

3.3.3 Soliton, collapse and turbulence

The outcome of the modulational instability depends on space dimensionality. The breakdown of a homogeneous state may lead all the way to small-scale fragmentation or the creation of singularities. Alternatively, stable finite-size objects may appear as an outcome of instability. As often happens, analysis of conservation laws helps to understand the destination of a complicated process.

Hamiltonian equation (3.41) conserves the energy (3.43). Since (3.41) describes wave propagation and four-wave scattering, it does not change the number of waves and thus conserves the wave action $N = \int |\psi|^2 \, d\mathbf{r}$. The conservation follows from the continuity equation

$$2i\partial_t |\psi|^2 = \omega'' \nabla (\psi^* \nabla \psi - \psi \nabla \psi^*) \equiv -2 \operatorname{div} \mathbf{J} . \qquad (3.44)$$

Note also the conservation of the momentum or total current, $\int \mathbf{J} \, d\mathbf{r}$, which does not play any role in this section but is important for Exercise 3.7. The symmetries responsible for the conservation of energy, momentum and action are respectively, invariance of (3.41) with respect to the time shift ($t \to t +$ const.), space shift ($\mathbf{r} \to \mathbf{r} +$ const.) and gauge invariance ($\psi \to \psi e^{i\alpha}$).

Consider a wave packet characterized by the generally time-dependent size l and the constant value of N.

Since one can estimate the typical value of the envelope in the packet as $|\psi|^2 \simeq N/l^d$, then $\mathcal{H} \simeq \omega''Nl^{-2} + TN^2l^{-d}$ – remember that the second term is negative here. We consider the conservative system, so the total energy is conserved yet we expect the radiation from the wave packet to bring it to the minimum of its energy. We wish to understand the direction of the evolution considering it adiabatically slow. Then, in the process of weak radiation, wave action is conserved since it is an adiabatic invariant. This is particularly clear for a quantum system, like a cloud of cold atoms, where N is their number. Whether this minimum corresponds to $l \to 0$ (which is called self-focusing or collapse) is determined by the balance between $|\nabla\psi|^2$ and $|\psi|^4$. The Hamiltonian $\mathcal{H}$ as a function of l in three different dimensionalities is shown in Figure 3.8.

(i) $d = 1$. At small l kinetic energy $\mathcal{H} \simeq \omega''Nl^{-2}$ dominates and leads to expansion, while attraction $\mathcal{H} \simeq -TN^2l^{-1}$ dominates at large l. It is thus clear that a stationary solution must exist with $l \sim \omega''/TN$, which minimizes the energy. Physically, the pressure of the waves balances the attraction force. Such a stationary solution is called a *soliton*, short for solitary wave. It is a traveling-wave solution of (3.41), $\psi(x,t) = A(x - ut)e^{i\varphi}$, with the amplitude function just moving and the phase having both a space-dependent traveling part and a uniform nonlinear part linearly growing with time: $\varphi(x,t) = f(x - ut) - TA_0^2t$. Here, complex A_0 and real u are soliton parameters. We substitute the travel solution into (3.41) and separate the real and imaginary parts:

$$A'' = \frac{2T}{\omega''}\left(A^3 - A_0^2A\right) + Af'\left(f' - \frac{2u}{\omega''}\right), \quad \omega''\left(A'f' + \frac{Af''}{2}\right) = uA'. \quad (3.45)$$

For the simple case of the standing wave ($u = 0$) the second equation gives $f = $ const., which can be put equal to zero. The first equation can be considered as a Newtonian equation $A'' = -dU/dA$ for the particle with coordinate A

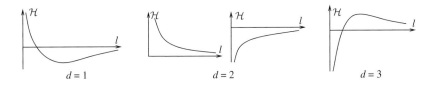

Figure 3.8 The Hamiltonian $\mathcal{H}$ as a function of the packet size l under fixed N.

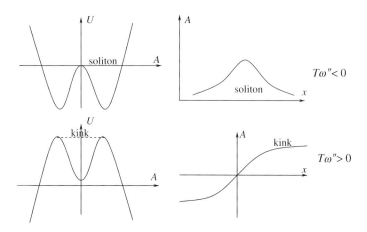

Figure 3.9 Energy as a function of the amplitude of a running wave, and the profile of the wave. The upper part corresponds to the case of an unstable condensate, where the wave is a soliton, the lower part to a stable condensate, where the wave is a kink.

in the potential $U(A) = -(T/2\omega'')(A^4 - 2A^2A_0^2)$ and the space coordinate x replacing the particle's time. The soliton is a separatrix, that is a solution that requires for particle an infinite time to reach zero, or in original terms where $A \to 0$ as $x \to \pm\infty$. The upper part of Figure 3.9 presumes $T/\omega'' < 0$, that is a case of modulational instability. Let me mention in passing that the separatrix also exists for $T/\omega'' > 0$ but in this case the running wave is a kink, that is a transition between two different values of the stable condensate (the lower part of the figure). The kink is seen as a dip in intensity $|\psi|^2$.

Considering a general case of a traveling soliton (at $T\omega'' < 0$), one can multiply the second equation by A and then integrate: $\omega''A^2 f' = u(A^2 - A_0^2)$, where by choosing the constant of integration we defined A_0 as A at the point where $f' = 0$. We can now substitute f' into the first equation and get the closed equation for A. The soliton solution has the form:

$$\psi(x,t) = \sqrt{2}A_0 \cosh^{-1}\left[\left(\frac{-2T}{\omega''}\right)^{1/2} A_0(x - ut)\right] e^{i(2x-ut)u/2\omega'' - iTA_0^2 t}.$$

Note that the Galilean transformation for the solutions of the NSE appears as $\psi(x,t) \to \psi(x - ut, t)\exp[iu(2x - ut)/2\omega'']$. In the original variable $a(\mathbf{r})$, our envelope solitons appear as shown in Figure 3.10.

(ii) $d = 2, 3$. When the condensate is stable, there exist stable solitons analogous to kinks, which are localized minima in the condensate intensity. In optics they can be seen as grey and dark filaments in a laser beam propagating through a nonlinear medium. The wave (condensate) amplitude turns into zero

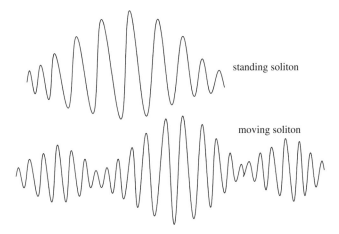

standing soliton

moving soliton

Figure 3.10 Standing and traveling solitons of the envelope of an almost monochromatic wave.

in a dark filament, which means that it is a vortex, i.e. a singularity of the wave phase, see Exercise 3.6.

When the condensate is unstable, there are no stable stationary solutions for $d = 2, 3$. From the dependence $\mathcal{H}(l)$ shown in Figure 3.8 we expect that the character of evolution will be completely determined by the sign of the Hamiltonian at $d = 2$: the wave packets with positive Hamiltonian spread because the wave dispersion (kinetic energy or pressure, in other words) dominates while the wave packets with negative Hamiltonian shrink and collapse. Let me stress that this way of arguing based on the dependence $\mathcal{H}(l)$ is nonrigorous and suggestive at best. A rigorous proof of the fact that the Hamiltonian sign determines whether the wave packet spreads or collapses in 2D is called Talanov's theorem, which is the expression for the second time derivative of the packet size squared, $l^2(t) = \int |\psi|^2 r^2 \, d\mathbf{r}$. To obtain that expression, differentiate over time using (3.44), then integrate by parts, then differentiate again:

$$
\frac{d^2 l^2}{\omega'' dt^2} = \frac{i\partial_t}{2} \int r^2 \nabla \left(\psi^* \nabla \psi - \psi \nabla \psi^* \right) d\mathbf{r} = i\partial_t \int r_\alpha \left(\psi \nabla_\alpha \psi^* - \psi^* \nabla_\alpha \psi \right) d\mathbf{r}
$$

$$
= - \int \left[\frac{\omega''}{2} \left(\psi \nabla_\alpha \Delta \psi^* + \psi^* \nabla_\alpha \Delta \psi - \nabla_\beta \left(\nabla_\beta \psi \nabla_\alpha \psi^* + \nabla_\beta \psi^* \nabla_\alpha \psi \right) \right) \right.
$$

$$
\left. - T \nabla_\alpha |\psi|^4 \right] r_\alpha \, d\mathbf{r} = dT \int |\psi|^4 \, d\mathbf{r} + 2\omega'' \int |\nabla \psi|^2 \, d\mathbf{r}
$$

$$
= 4\mathcal{H} + 2(d-2)T \int |\psi|^4 \, d\mathbf{r}.
$$

Consider an unstable case with $T\omega'' < 0$. We see that indeed for $d \geq 2$ one has an inequality $\partial_{tt} l^2 \leq 4\omega'' \mathcal{H}$ so that

$$l^2(t) \leq 2\omega'' \mathcal{H} t^2 + C_1 t + C_2$$

and for $\omega'' \mathcal{H} < 0$ the packet shrinks to singularity in a finite time (this is the singularity in the framework of NSE, which is itself valid only for the scales much larger than the wavelength of the carrier wave $2\pi/k_0$). This, in particular, describes self-focusing of light in nonlinear media. For $d = 2$ and $\omega'' \mathcal{H} > 0$, on the contrary, one has dispersive expansion and decay.

Turbulence with two cascades. As mentioned, any equation (3.40) that describes only four-wave scattering necessarily conserves both the energy $\mathcal{H}$ and the number of waves (or wave action) N. For waves of small amplitude, the energy is approximately quadratic in wave amplitudes, $\mathcal{H} \approx \int \omega_k |a_k|^2 \, d\mathbf{k}$, as well as $N = \int |a_k|^2 \, d\mathbf{k}$. The existence of two quadratic positive integrals of motion in a closed system means that if such system is subject to external pumping and dissipation, it may develop turbulence consisting of two cascades.

Indeed, imagine that the source at some ω_2 pumps N_2 waves per unit time. It is then clear that for a steady state one needs two dissipation regions in ω-space (at some ω_1 and ω_3) to absorb the inputs of both N and E. Conservation laws allow one to determine the numbers of waves, N_1 and N_3, absorbed per unit time in the regions of low and high frequencies, respectively. Schematically, solving $N_1 + N_3 = N_2$ and $\omega_1 N_1 + \omega_3 N_3 = \omega_2 N_2$ we get

$$N_1 = N_2 \frac{\omega_3 - \omega_2}{\omega_3 - \omega_1}, \qquad N_3 = N_2 \frac{\omega_2 - \omega_1}{\omega_3 - \omega_1}. \tag{3.46}$$

We see that for a sufficiently large left interval (when $\omega_1 \ll \omega_2 < \omega_3$) most of the energy is absorbed by the right sink: $\omega_2 N_2 \approx \omega_3 N_3$. Similarly at $\omega_1 < \omega_2 \ll \omega_3$ most of the wave action is absorbed at small ω: $N_2 \approx N_1$. When $\omega_1 \ll \omega_2 \ll \omega_3$ we have two cascades with the fluxes of energy ε and wave action Q. The Q-cascade toward large scales is called the inverse cascade (Kraichnan 1967, Zakharov 1967); it corresponds, somewhat counter-intuitively, to a kind of self-organization, i.e. the creation of larger and slower modes out of small-scale fast fluctuations.[9] The limit $\omega_1 \to 0$ is well-defined; in this case the role of the left sink can actually be played by a condensate, which absorbs an inverse cascade. Note in passing that consideration of thermal equilibrium in a finite-size system with two integrals of motion leads to the notion of negative temperature.[10]

An important hydrodynamic system with two quadratic integrals of motion is a two-dimensional incompressible ideal fluid. In two dimensions, the velocity $\mathbf{u}$ is perpendicular to the vorticity $\omega = \nabla \times \mathbf{u}$, so that the vorticity of any fluid element is conserved by virtue of the Kelvin theorem. This means that the space integral of any function of vorticity is conserved, including $\int \omega^2 \, d\mathbf{r}$, called enstrophy. We can write the densities of the two quadratic integrals of motion, energy and enstrophy, in terms of the velocity spectral density: $E = \int |\mathbf{v_k}|^2 \, d\mathbf{k}$ and $\Omega = \int |\mathbf{k} \times \mathbf{v_k}|^2 \, d\mathbf{k}$. Assume now that we excite turbulence with a force having a wavenumber k_2 while dissipation regions are at k_1, k_3. Applying the consideration similar to (3.46), we express the energy dissipation rates E_1, E_3 via the input rate E_2:

$$E_1 = E_2 \frac{k_3^2 - k_2^2}{k_3^2 - k_1^2}, \qquad E_3 = E_2 \frac{k_2^2 - k_1^2}{k_3^2 - k_1^2}. \tag{3.47}$$

We see that for $k_1 \ll k_2 \ll k_3$, most of the energy is absorbed by the left sink, $E_1 \approx E_2$, and most of the enstrophy is absorbed by the right one, $\Omega_2 = k_2^2 E_2 \approx \Omega_3 = k_3^2 E_3$. We conclude that conservation of both energy and enstrophy in two-dimensional flows requires two cascades: that of the enstrophy toward small scales and that of the energy toward large scales (opposite to the direction of the energy cascade in three dimensions). Large-scale motions of the ocean and planetary atmospheres can be considered to be approximately two-dimensional; the creation and persistence of large-scale flow patterns in these systems is probably related to inverse cascades.[11]

3.4 Korteveg–de-Vries (KdV) equation

Here we consider another universal limit: weakly nonlinear long waves. In the long-wave limit one may expand frequency in the powers of wavenumber (more accurately, reversibility of a Hamiltonian system means that we expand ω^2 in powers of k^2). If homogeneous perturbation ($k = 0$) does not cost any energy then this expansion starts from the first term, that is the dispersion relation of such waves is close to acoustic. We derive the respective KdV equation for shallow-water waves. We then consider some remarkable properties of this equation and of such waves.

3.4.1 Waves in shallow water

Linear gravity-capillary waves have $\omega_k^2 = (gk + \alpha k^3/\rho)\tanh kh$, see (3.17). That is, for sufficiently long waves (when the wavelength is larger than both h and $\sqrt{\alpha/\rho g}$) their dispersion relation is close to linear:

$$\omega_k = \sqrt{gh}\,k - \beta k^3, \qquad \beta = \frac{\sqrt{gh}}{2}\left(\frac{h^2}{3} - \frac{\alpha}{\rho g}\right). \tag{3.48}$$

That means that shallow-water waves are similar to acoustic waves with the speed of sound $c = \sqrt{gh}$. In the nineteenth century, J. S. Russel used this formula to estimate the atmosphere height, observing propagation of weather changes, that is of pressure waves. Taking $h \simeq 10$ km, we obtain $c \simeq 320$ m s^{-1}. Non-surprisingly, it is comparable with the speed of sound in the air near the surface, $c = \sqrt{\gamma P/\rho}$, with $P \simeq \rho gh$.

When different harmonics propagate with almost the same velocity, one can expect a quasi-simple plane wave propagating in one direction, like that described in Sections 2.3.2 and 2.3.3. The main effect will be propagation with the speed $\sqrt{gh}$ without changing form, while small effects of nonlinearity and dispersion will lead to slow changes. Let us derive the equation describing such a wave. From the dispersion relation, we obtain the linear part of the equation: $u_t + \sqrt{gh}u_x = -\beta u_{xxx}$ or in the reference frame moving with the velocity $\sqrt{gh}$ one has $u_t = -\beta u_{xxx}$. To derive the nonlinear part of the equation in the long-wave limit, it is enough to consider the first nonvanishing spatial derivative (i.e. the first). The motion is close to one-dimensional so that $u = v_x \gg v_z$. The z component of the Euler equation gives $\partial p/\partial z = -\rho g$ and $p = p_0 + \rho g(\zeta - z)$. Here $\zeta(x,t)$ is again the elevation of the surface, where the pressure is assumed to be p_0. Now we substitute the pressure into the x component of the Euler equation:

$$\frac{\partial u}{\partial t} + u\frac{\partial u}{\partial x} = -\frac{1}{\rho}\frac{\partial p}{\partial x} = -g\frac{\partial \zeta}{\partial x}.$$

In the continuity equation, $h + \zeta$ now plays the role of density:

$$\frac{\partial \zeta}{\partial t} + \frac{\partial}{\partial x}(h + \zeta)u = 0.$$

We differentiate it with respect to time, substitute the Euler equation and neglect the cubic term:

$$\frac{\partial^2 \zeta}{\partial t^2} = \partial_x(h + \zeta)(uu_x + g\zeta_x) + \partial_x u \partial_x(h + \zeta)u$$

$$\approx \frac{\partial^2}{\partial x^2}\left(gh\zeta + \frac{g}{2}\zeta^2 + hu^2\right). \tag{3.49}$$

Apparently, the right-hand side of this equation contains terms of different orders. The first term describes the linear propagation with velocity $\sqrt{gh}$ while the rest describe a small nonlinear effect. Such equations are usually treated by the method called multiple time (or multiple scale) expansion. We actually applied this method in deriving the Burgers and nonlinear Schrödinger equations. We assume that u and ζ depend on two arguments, namely $u(x - \sqrt{gh}t,t)$, $\zeta(x - \sqrt{gh}t,t)$, and, that the dependence on the second argument is slow. In what follows, we write the equation for u. Then at the main order $\partial_t u = -\sqrt{gh}u_x = -g\zeta_x$ that is $\zeta = u\sqrt{h/g}$, which is a direct analogue of $\delta\rho/\rho = u/c$ for acoustics. From now on, u_t denotes the derivative with respect to the slow time (or simply speaking, in the reference frame moving with velocity $\sqrt{gh}$). We obtain from (3.49):

$$(\partial_t - \sqrt{gh}\partial_x)(\partial_t + \sqrt{gh}\partial_x)u \approx -2\sqrt{gh}u_{xt} = (3/2)\sqrt{gh}(u^2)_{xx},$$

that is the nonlinear contribution into u_t is $-3uu_x/2$. Comparing this with the general acoustic expression (2.34), which is $-(\gamma+1)uu_x/2$, we see that shallow-water waves correspond to $\gamma = 2$. This is also clear from the fact that the local "sound" velocity is $\sqrt{g(h+\zeta)} \approx \sqrt{gh} + u/2 = c + (\gamma-1)u/2$, see (2.33).

The analogy between shallow-water waves and sound means that there exist shallow-water shocks called bores[12] and hydraulic jumps. The Froude number u^2/gh plays the role of the (squared) Mach number in this case.

Hydraulic jumps can be readily observed in the kitchen sink when water from the tap spreads radially with the speed, exceeding the linear "sound" velocity $\sqrt{gh}$: the fluid layer thickness suddenly increases, which corresponds to a shock, see Figure 3.11 and Exercise 3.8. This shock is sent back by the sink sides that stop the flow; the jump position is where the shock speed is equal to the flow velocity.[13] For a weak shock, the shock speed is the speed of "sound" so that the flow is "supersonic" inside and "subsonic" outside. Long surface waves cannot propagate into the interior region, which can thus be called a white hole (as opposed to a black hole in general relativity) with the hydraulic jump playing the role of a horizon. In Figure 3.11 one sees circular capillary ripples propagating inside; these are to be distinguished from the jump itself which is noncircular.

3.4.2 The KdV equation and the soliton

Now we are ready to combine both the linear term from the dispersion relation and the nonlinear term just derived. Making a change $u \to 2u/3$ we turn the coefficient at the nonlinear term into unity. The equation

Figure 3.11 Hydraulic jump in a kitchen sink. Photograph copyright: Joe Gough, www.dreamstime.com.

$$u_t + u u_x + \beta u_{xxx} = 0 \qquad\qquad (3.50)$$

has been derived by Korteveg and de Vries in 1895 and is called the KdV equation. Together with the nonlinear Schrödinger and Burgers equations, it is a member of an exclusive family of universal nonlinear models. It is one-dimensional, like the Burgers equation, and has the same degree of universality. Namely, the systems with a continuous symmetry (say, translation invariance) spontaneously broken allow for what is called a Goldstone mode with $\omega \to 0$ when $k \to 0$. When wavelength is larger than any other scale in conservative time-reversible center-symmetrical systems, one expects an acoustic branch of excitations with an analytic function $\omega^2(k^2)$. That function has the expansion in integer powers, k^2, k^4, etc[14]. For long waves moving in one direction one has $\omega_k = ck[1 + C(kl_0)^2]$, where C is a dimensionless coefficient of order unity and l_0 is some internal scale in the system. For gravity water waves, l_0 is the water depth; for capillary waves, it is $\sqrt{\alpha/\rho g}$. We see from (3.48) that, depending on which scale is larger, β can be either positive or negative, which corresponds to the waves with finite k moving slower or faster, respectively, than the 'sound' velocity $\sqrt{gh}$. Indeed, adding surface tension to the restoring force, one increases the frequency. On the other hand, the finite depth-to-wavelength ratio means that fluid particles move in ellipses (rather than in straight lines) which decreases the frequency. Quadratic nonlinearity $\partial_x u^2$ (which occurs in both the Burgers and KdV equations and means effectively the renormalization of the speed of sound) is also pretty general, indeed, it has to be zero for a uniform velocity so that it contains a derivative. An incomplete list of the excitations described by the KdV equation contains acoustic perturbations in a plasma (where l is either the Debye radius of charge screening or the Larmor cyclotron radius in the magnetized plasma), phonons in solids (where l is the distance between atoms) and phonons in helium (in this case, amazingly, the sign of β depends on pressure).

The KdV equation, has a symmetry $\beta \rightarrow -\beta$, $u \rightarrow -u$ and $x \rightarrow -x$, which makes it enough to consider only positive β. Let us first look for the traveling waves, that is substitute $u(x - vt)$ into (3.50):

$$\beta u_{xxx} = vu_x - uu_x.$$

This equation has a symmetry $u \rightarrow u + w$, $v \rightarrow v + w$ so that integrating it once we can set the integration constant to zero by choosing an appropriate constant w (this is the trivial renormalization of sound velocity due to a uniformly moving fluid). We thus get the equation

$$\beta u_{xx} = -\frac{\partial U}{\partial u}, \qquad U(u) = \frac{u^3}{6} - \frac{vu^2}{2} + \text{const.}$$

A general solution of this ordinary differential equation can be written in elliptic functions. We don't need it though to understand the general properties of the solutions and to pick up the special one (i.e. the soliton). Just like in Section 3.3.3, we can treat this equation as a Newtonian equation for a particle in a potential, treating u as a particle coordinate and x as time. We consider positive v since it is a matter of choosing a proper reference frame. Since unrestricted growth would violate the assumption of weak nonlinearity, we restrict ourselves by the solution with finite $|u(x)|$. Such finite solutions have u in the interval $(0, 3v)$.

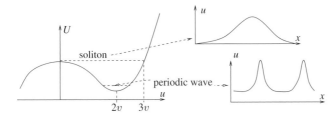

We see that quasilinear periodic waves exist near the bottom at $u \approx 2v$. Their amplitude is small in the reference frame moving with $-2v$ – in that reference frame they have negative velocity as must be the case for positive β. Indeed, the sign of β is minus the sign of the dispersive correction $-3\beta k^2$ to the group velocity $d\omega_k/dk$. The soliton, on the contrary, moves with positive velocity in the reference frame where there is no perturbation at infinity (it is precisely the reference frame used in the picture). Solitons are supersonic if the periodic waves are subsonic and vice versa (the physical reason for this is that the soliton should not be able to radiate resonant linear waves).

As usual, the soliton solution is a separatrix passing through $u = 0$:

$$u(x,t) = 3v \cosh^{-2}\left[\sqrt{\frac{v}{4\beta}}\,(x - vt)\right]. \tag{3.51}$$

The higher the amplitude, the faster it moves (for $\beta > 0$) and the more narrow it is. Like the argument at the end of Section 3.3.2, one can realize that the 1D KdV soliton is unstable with respect to the perpendicular perturbations if its propagation speed $\sqrt{gh} + v$ decreases with the amplitude, that is for $\beta < 0$ when $v < 0$ in (3.51), so that the soliton is subsonic and linear waves are supersonic.

Without friction or dispersion, nonlinearity breaks acoustic perturbation. We see that wave dispersion stabilizes the wave like the viscous friction stabilizes the shock front, but the waveform is, of course, different. The ratio of nonlinearity to dispersion, $\sigma = uu_x/\beta u_{xxx} \sim ul^2/\beta$, is an intrinsic nonlinearity parameter within the KdV equation (in addition to the original "external" nonlinearity parameter, the Mach number u/c, assumed to be always small). For a soliton, $\sigma \simeq 1$, that is nonlinearity balances dispersion as was the case with the NSE soliton from Section 3.3.3. This also shows that the soliton is a nonperturbative object; one cannot derive it by starting from a linear traveling waves and treating nonlinearity perturbatively. For the Burgers equation, both finite-M and traveling-wave solutions depended smoothly on the respective intrinsic parameter Re and existed for any Re. Introduction of σ leads to natural questions: Can we assert that any perturbation with $\sigma \ll 1$ corresponds to linear waves? What can one say about the evolution of a perturbation with $\sigma \gg 1$ within the KdV equation?

3.4.3 Inverse scattering transform

It is truly remarkable that the evolution of arbitrary initial perturbation can be studied analytically within the KdV equation. It is done in a somewhat unexpected way by considering a linear stationary Schrödinger equation with the function $-u(x,t)/6\beta$ as a potential, depending on time as a parameter:

$$\left[-\frac{d^2}{dx^2} - \frac{u(x,t)}{6\beta}\right]\Psi = E\Psi. \tag{3.52}$$

Positive $u(x)$ could create bound states, that is a discrete spectrum. As has been noticed by Gardner, Green, Kruskal and Miura in 1967, the spectrum

Figure 3.12 Asymptotic form of a localized perturbation.

E does not depend on time if $u(x,t)$ evolves according to the KdV equation. In other words, the KdV equation describes the iso-spectral transformation of the quantum potential. To show this, express u via Ψ

$$u = -6\beta \left(E + \frac{\Psi_{xx}}{\Psi} \right). \tag{3.53}$$

Notice the similarity to the Hopf substitution one uses for the Burgers equation, $v = -2v\phi_\xi/\phi$. There is one more derivative in (3.53) because there is one more derivative in the KdV equation – despite the seemingly naive and heuristic nature of such thinking, it is precisely the method by which Miura came to suggest (3.53). Now, substitute (3.53) into the KdV equation and derive

$$\Psi^2 \frac{dE}{dt} = 6\beta \partial_x \left[(\Psi \partial_x - \Psi_x)(\Psi_t + \Psi_{xxx} - \Psi_x(u+E)/2) \right]. \tag{3.54}$$

Integrating it over x we get $dE/dt = 0$ since $\int_{-\infty}^{\infty} \Psi^2 \, dx$ is finite for a bound state. The eigenfunctions evolve according to the equation that can be obtained by twice integrating (3.54) and setting the integration constant to zero because of the normalization:

$$\Psi_t + \Psi_{xxx} - \Psi_x \frac{u+E}{2} = 0. \tag{3.55}$$

From the viewpoint of (3.52), the soliton is the well with exactly one level, $E = v/8\beta$, which could be checked directly. For distant solitons, one can define energy levels independently. For different solitons, velocities are different and they will generally have collisions. Since the spectrum is conserved, after all the collisions we have to have the same solitons. Since the velocity is proportional to the amplitude, the final state of the perturbation with $\sigma \gg 1$ must look like a linearly ordered sequence of solitons: the quasilinear waves that correspond to a continuous spectrum are left behind and eventually spread. One can prove this and analyze the evolution of an arbitrary initial perturbation using the *inverse scattering transform* (IST) method:

$$u(x,0) \to \Psi(x,0) = \sum a_n \Psi_n(x,0) + \int a_k \Psi_k(x,0)\,\mathrm{d}k$$
$$\to \Psi(x,t) = \sum a_n \Psi_n(x,t) + \int a_k \Psi_k(x,0)\mathrm{e}^{-\mathrm{i}\omega_k t}\,\mathrm{d}k \qquad (3.56)$$
$$\to u(x,t).$$

The first step is to find the eigenfunctions and eigenvalues in the potential $u(x,0)$. The second (trivial) step is to evolve the discrete eigenfunctions according to (3.55) and the continuous eigenfunctions according to frequency. The third (nontrivial) step is to solve an inverse scattering problem, that is to restore the potential $u(x,t)$ from its known spectrum and the set of (new) eigenfunctions.[15]

Considering weakly nonlinear initial data ($\sigma \ll 1$), one can treat the potential energy as a perturbation in (3.52) and use the results from the quantum mechanics of shallow wells. Remember that the bound state exists in 1D if the integral of the potential is negative, that is in our case when the momentum $\int u(x)\,\mathrm{d}x > 0$ is positive, i.e. the perturbation is supersonic. The subsonic small perturbation (with a negative momentum) does not produce solitons, it produces subsonic quasi-linear waves. On the other hand, however small the initial nonlinearity σ is, the soliton (an object with $\sigma \sim 1$) will necessarily appear in the course of evolution of the perturbation with positive momentum. The amplitude of the soliton is proportional to the energy of the bound state, which is known to be proportional to $(\int u\,\mathrm{d}x)^2$ in the shallow well.

The same IST method was applied to the 1D nonlinear Schrödinger equation by Zakharov and Shabat in 1971. Now, the eigenvalues E of the system

$$\mathrm{i}\partial_t \psi_1 + \psi \psi_2 = E \psi_1$$
$$-\mathrm{i}\partial_t \psi_2 - \psi^* \psi_1 = E \psi_2$$

are conserved when ψ evolves according to the 1D NSE. Like the KdV equation, one can show that within the NSE, an arbitrary localized perturbation evolves into a set of solitons and a diffusing quasilinear wave packet.

The reason why universal dynamic equations in one space dimension happen to be integrable may be related to their universality. Indeed, in many different classes of systems, weakly nonlinear long-wave perturbations are described by the Burgers or KdV equations and quasimonochromatic perturbations by the NSE. Those classes may happen to contain degenerate integrable cases; then integrability exists for the limiting equations as well. These systems are actually two-dimensional (space plus time) and their integrability can be related to a unique role of the conformal group (which is infinite in 2D). From the potential flows described in Section 1.2.4 to the nonlinear waves described in this chapter, it seems that the complex analysis and the idea of analyticity is behind most of the solvable cases in fluid mechanics, as well as in other fields of physics.

Exercises

3.1 Why is it that a beachcomber (a long water wave rolling upon the beach) usually comes to the coast being almost parallel to the coastline even when the wind blows at an angle?

3.2 A quasimonochromatic packet of waves contains N crests and wells propagating along the fluid surface. How many "up and down" motions does a light float undergo while the packet passes? Consider two cases: (i) gravity waves on deep water, (ii) capillary waves on deep water.

3.3 Dropping a stone into the deep water, one could see, after a little while, waves propagating outside an expanding circle of fluid at rest. Sketch a snapshot of the wave crests. What is the velocity of the boundary of the quiescent fluid circle?

3.4 The existence of the stable small-amplitude waves that are described by (3.32) cannot be taken for granted. Consider a general form of the quadratic Hamiltonian

$$\mathscr{H}_2 = \int \left[A(k)|b_k|^2 + B(k)(b_k b_{-k} + b_k^* b_{-k}^*) \right] d\mathbf{k}. \qquad (3.57)$$

(i) Find the linear transformation (the Bogoliubov u–v transformation) $b_k = u_k a_k + v_k a_{-k}^*$ that turns (3.57) into (3.32).

(ii) Consider the case when the even part $A(k) + A(-k)$ changes sign on some surface (or line) in k-space while $B(k) \neq 0$ there. What does this mean physically? In this case, what is the simplest form that the quadratic Hamiltonian $\mathscr{H}_2$ can be turned into?

3.5 Show that the power dispersion relation $\omega_k \propto k^\alpha$ is of the decay type if $\alpha \geq 1$, i.e. it is possible to find such $\mathbf{k}_1, \mathbf{k}_2$ that $\omega(\mathbf{k}_1 + \mathbf{k}_2) = \omega_1 + \omega_2$. Consider two-dimensional space $\mathbf{k} = \{k_x, k_y\}$. Hint: ω_k is a concave surface and the resonance condition can be thought of as an intersection of some two surfaces.

3.6 Consider the case of a stable condensate in 3D and describe a solution of the NSE equation (3.41) having the form

$$\psi = A e^{-iTA^2 t + i\phi} f(r/r_0),$$

where r is the distance from the vortex axis and ϕ is a polar angle. Are the parameters A and r_0 independent? Describe the asymptotics of f for small and large distances. Why is it called a vortex?

3.7 Consider a discrete spectral representation of the Hamiltonian of the 1D NSE in a finite medium:

$$\mathscr{H} = \sum_m \beta m^2 |a_m|^2 + (T/2) \sum_{ikm} a_i a_k a_{i+k-m}^* a_m^*.$$

Take only three modes $m = 0, 1, -1$. Describe the dynamics of such a three-mode system.

3.8 Calculate the energy dissipation rate per unit length of the hydraulic jump. Fluid flows with the velocity u_1 in a thin layer of height h_1 such that the Froude number slightly exceeds unity: $u_1^2/gh_1 = 1 + \varepsilon$, $\varepsilon \ll 1$.

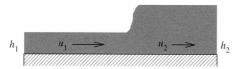

3.9 If one ought to take into account both dissipation and dispersion of a sound wave, then the so-called KdV–Burgers equation arises:

$$u_t + u u_x + \beta u_{xxx} - \mu u_{xx} = 0. \tag{3.58}$$

For example, it allows one to describe the influence of dispersion on the structure of a weak shock wave. Consider a traveling solution $u_0(x - vt)$ of this equation, assuming zero conditions at $+\infty$: $u_0 = u_{0x} = u_{0xx} = 0$. Sketch the form of $u_0(x)$ for $\mu \ll \sqrt{\beta v}$ and for $\mu \gg \sqrt{\beta v}$.

3.10 Find the dispersion relation of the waves on the boundary between two fluids, one flowing and one still, in the presence of gravity g and the surface tension α. Describe possible instabilities. Consider, in particular, the cases $\rho_1 > \rho_2$, $v = 0$ (inverted gravity) and $\rho_1 \ll \rho_2$ (wind upon water).

$$
\begin{array}{ll}
v_1 = v & \rho_1 \\
\hline
v_2 = 0 & \rho_2
\end{array}
\quad \alpha \quad \Big| \; g
$$

3.11 We have seen in Sections 2.3.1 and 3.1.1 that a propagating wave can induce net fluid flow called Stokes drift. Apparently, a standing wave cannot induce a drift. But what if we have a velocity field as a linear superposition of *two* standing waves: $\mathbf{v}(\mathbf{r},t) = \sin(\omega t)\mathbf{V}(\mathbf{r}) + \sin(\omega t + \varphi)\mathbf{U}(\mathbf{r})$? Velocity in every point still averages to zero. Integrate the trajectory of a fluid particle over the period and find out if the net drift is possible. Assume for simplicity that the velocity gradients are much smaller than ω.

4

Solutions to exercises

A lucky guess is never merely luck.

(Jane Austen)

4.1 Chapter 1

1.1 Consider a prism inside a fluid.

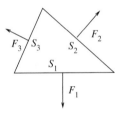

The forces must sum into zero in equilibrium, which means that after being rotated by $\pi/2$, the force vectors form a closed triangle similar to that of the prism. Therefore, the forces are proportional to the areas of the respective faces and the pressures are equal. Note that we do not assume isotropy of fluid properties.

In a moving fluid, the difference between the diagonal components of the stress tensor must be zero for a uniform flow and is therefore proportional to velocity gradients as discussed in Section 1.4.2.

The fact that fluids transfer pressure equally in all directions is used in many hydraulic devices including the braking system of most motor vehicles. It also allows for force amplification. Indeed, consider liquid in a U tube closed by pistons in the limbs of different areas, S_1 and S_2. Upon application of a force F_1 to the small piston, it exerts against the liquid the pressure F_1/S_1 that is equal to the pressure the liquid exerts against the large piston, which thus is able to apply larger force: $F_2 = F_1 S_2/S_1$.

1.2 The gravitational force must balance the gradient of pressure: $-\rho\nabla\phi = \nabla p$. Dividing it by ρ, acting by div and using $\Delta\phi = 4\pi G\rho$, one obtains

$$\frac{d}{dr}\left(\frac{r^2 dp}{\rho dr}\right) = -4\pi Gr^2\rho \ . \tag{4.1}$$

For an incompressible liquid, one finds

$$p = \frac{2}{3}\pi G\rho^2(R^2 - r^2) \ .$$

Since the pressure turns into zero at the boundary, then a self-gravitating body of liquid can exist in a vacuum.

For an ideal gas under a constant temperature, one uses $\rho = p/c^2$ and obtains

$$\frac{d}{dr}\left(\frac{r^2 dp}{pdr}\right) = -4\pi Gc^{-4}r^2 p \ . \tag{4.2}$$

It is straightforward to obtain a particular exact solution of this nonlinear equation:

$$p(r) = \rho(r)c^2 = c^4/2\pi Gr^2 \ . \tag{4.3}$$

This solution does not contain any constant of integration, it diverges both at zero (infinite pressure) and at infinity (infinite total mass). Finite pressure at the center, p_0, sets the core radius $r_0 = c^2/(Gp_0)^{1/2}$, so that $p(r) \approx p_0(1 - 2Gp_0 r^2/3c^4)$ at $r \ll r_0$. At $r \gg r_0$, the physical solution approaches the singular one (4.3) and continues until some finite outer radius set by a finite total mass. At a finite radius, the inside pressure is finite and it requires some external pressure (apparently less than p_0) to balance it. Pressure values at the center and at the boundary are two constants that define the solution of this differential equation of the second order.

For a general adiabatic relation $p \propto \rho^\gamma$, the power asymptotic is $\rho \propto r^{2/(\gamma-2)}$ but it is realized only for $\gamma > 4/3$ when it is integrable at infinity. For $1 < \gamma < 4/3$, the integrable asymptotic is $\rho \propto r^{1/(1-\gamma)}$; in particular, for $\gamma = 6/5$ there is a simple solution $p \propto \rho^{6/5} \propto (a^2 + r^2)^{-3}$ valid and regular everywhere and having a finite mass.

1.3 The discharge rate is $S'\sqrt{2gh}$. Energy conservation gives us $v = \sqrt{2gh}$ at the vena contracta. This velocity squared times the density times the area S' gives the horizontal momentum flux, which must be determined by the force exerted by the walls on the fluid. For the tube, projecting inward (called Borda's mouthpiece), one can neglect the motion near the walls so that pressures on the opposite walls are equal everywhere; the only unbalanced part is opposite the hole. The force imbalance is the pressure $p = \rho g h$ times the hole area S. We thus get $\rho v^2 S' = \rho g h S$ and $S' = S/2$. For the orifice, drilled directly in the wall as in Figure 1.3, one cannot neglect the motion near the wall. Generally, that motion diminishes the pressure near the exit and makes the force imbalance larger. The reaction force is therefore greater and so is the momentum flux. Since the jet exits with the same velocity it must have a larger cross-section, so that $S'/S \geq 1/2$ (for a round hole in a thin wall it is known empirically that $S'/S \simeq 0.62$). To maximize the discharge rate for a given exit area, Borda suggested to attach a tube inside with twice the area (and length exceeding the diameter).

This general argument based on the conservation laws of energy and momentum works in any dimension. For Borda's mouthpiece in two dimensions, one can describe the whole flow neglecting gravity and assuming the (plane) tube infinite; both assumptions are valid for a flow not far from the corner. This is not the symmetric flow shown in Figure 1.17 with $n = 1/2$, which has an infinite velocity at the edge. We need a flow along the tube wall on the one side and detached from it on the other side. The asymmetric flow with separation also can be described using conformal maps but now with singularities, as shown in the following figure:

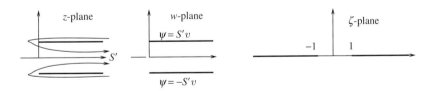

The tube walls coincide with the streamlines. Separation means jumps in the potential on the walls, which requires cuts in the ζ-plane. This corresponds to the potential $w \propto \ln \zeta$ with the velocity finite everywhere; the details can be found in Section 11.51 of [20]. For a slit in a thin wall, a 2D solution can be found in Section 11.53 of [20] or Chapter 10 of [16], which gives the coefficient of contraction $\pi/(\pi+2) \approx 0.61$.

1.4 Simply speaking, vorticity is the velocity circulation ($=$ vorticity flux) divided by the area while the angular velocity Ω is the velocity circulation (around the particle) divided by the radius a and the circumference $2\pi a$:

$$\Omega = \int u\,\mathrm{d}l/2\pi a^2 = \int \omega\,\mathrm{d}f/2\pi a^2 = \omega/2 .$$

One can obtain the same result a bit more formally: place the origin inside the particle and consider the velocity of a point of the piece with radius vector $\mathbf{r}$. Since the particle is small we use the Taylor expansion $v_i(\mathbf{r}) = S_{ij}r_j + A_{ij}r_j$ where $S_{ij} = (\partial_i v_j + \partial_j v_i)/2$ and $A_{ij} = (\partial_i v_j - \partial_j v_i)/2$. A rigid body cannot be deformed, thus $S_{ij} = 0$. The only isometries that do not deform the body and do not shift its centre of mass are the rotations. The rotation with the angular velocity Ω gives $A_{ik}r_k = \varepsilon_{ijk}\Omega_j r_k$. On the other hand, the vorticity component is

$$\omega_i \equiv [\nabla \times \mathbf{v}]_i = \varepsilon_{ijk}\partial_j v_k = \frac{1}{2}\varepsilon_{ijk}(\partial_j v_k - \partial_k v_j).$$

Using the identity $\varepsilon_{ijk}\varepsilon_{imn} = \delta_{jm}\delta_{kn} - \delta_{jn}\delta_{mk}$ and

$$\varepsilon_{imn}\omega_i = \frac{1}{2}(\delta_{jm}\delta_{kn} - \delta_{jn}\delta_{mk})(\partial_j v_k - \partial_k v_j) = \partial_m v_n - \partial_n v_m ,$$

we derive $A_{ik}r_k = \varepsilon_{ijk}\omega_j r_k/2$ and $\Omega = \omega/2$.

1.5 Use the Bernoulli equation, written for the point of maximal elevation (when $v = 0$ and the height is H) and at infinity: $2gH = 2gh + v_\infty^2$.

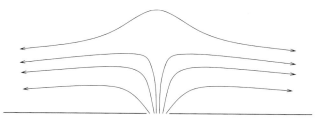

   (i)   The flow is two-dimensional and far from the slit has only a horizontal velocity, which does not depend on the vertical coordinate because of potentiality. Conservation of mass requires $v_\infty = q/2\rho h$ and the elevation $H - h = q^2/8g\rho^2 h^2$.

 (ii)   There is no elevation for a potential flow in this case since the velocity goes to zero at large distances (as an inverse distance from the source). A fountain with an underwater source surely results from a nonpotential flow.

1.6 Streamlines by definition are the lines where

$$\mathbf{v} \times \mathrm{d}\mathbf{l} = \left(v_r\hat{\mathbf{r}} + v_\theta\hat{\theta}\right) \times \left(\mathrm{d}r\hat{\mathbf{r}} + \mathrm{d}\theta\,\hat{\theta}\hat{z}\right) = rv_r\mathrm{d}\theta - v_\theta\mathrm{d}r = 0,$$

so the relation between r and θ is defined by the equation

$$\frac{d\theta}{dr} = \frac{v_\theta}{rv_r}. \tag{4.4}$$

In the reference frame where the fluid is at rest at infinity,

$$v_r = -u\cos\theta \, (R/r)^3, \quad v_\theta = -(1/2)u\sin\theta \, (R/r)^3. \tag{4.5}$$

Integrating

$$\frac{d\theta}{dr} = \frac{\tan\theta}{2r},$$

one obtains $r \propto \sin^2\theta$. Every such line starts at $r=0, \theta=0$, reaches its maximal radius for $\theta = \pi/2$, and returns to $r \to 0$ as $\theta \to \pi$; of course, only parts for $r > R$ describe streamlines shown in the left part of Figure 4.1.

Because of axial symmetry, the flow past a sphere is actually a set of plane flows, so that one can introduce the stream function, analogous to that in two dimensions. In a plane through the axis of symmetry, the difference of the stream function between two points is the flux through the annular surface generated by any line between those two points. For the infinitesimal vector ds in the plane, such flux is $2\pi d\psi = 2\pi r$ $\sin\theta \mathbf{v} \times \mathbf{dl}$, so that, for instance, $v_\theta = (r\sin\theta)^{-1}\partial\psi/\partial r$. For (4.5) we obtain $\psi = -uR^3 \sin^2\theta/2r$. The streamlines are level lines of the stream function. If one defines a vector whose only component is perpendicular to the plane and equal to the stream function then the velocity is the curl of that vector in cylindrical coordinates.

In the reference frame of the sphere, the velocity of the inviscid potential flow is as follows:

$$v_r = u\cos\theta \left(1 - (R/r)^3\right),$$

$$v_\theta = -u\sin\theta \left(1 + (R^3/2r^3)\right).$$

The streamlines are defined by the equation

$$\frac{d\theta}{dr} = -\frac{2r^3 + R^3}{2r(r^3 - R^3)}\tan\theta,$$

whose integration gives the streamlines

$$-\int_{\theta_1}^{\theta_2} d\theta \frac{2}{\tan\theta} = \int_{r_1}^{r_2} dr \frac{2r^3 + R^3}{r(r^3 - R^3)}$$

$$\left(\frac{\sin\theta_1}{\sin\theta_2}\right)^2 = \frac{r_2(r_1 - R)\left(r_1^2 + r_1 R + R^2\right)}{r_1(r_2 - R)\left(r_2^2 + r_2 R + R^2\right)}.$$

It corresponds to the stream function in the sphere reference frame as follows: $\psi = -u\sin^2\theta(r^2 - R^3/r)/2$. The streamlines relative to the sphere are in the right part of Figure 4.1. For two-dimensional case, see Section 9.20–22 of [20].

From the velocity of the viscous Stokes flow given by (1.59), one can obtain the stream function. In the reference frame where the fluid is at rest at infinity, $\psi = urR\sin^2\theta(3/4 - R^3/3r^3)$, the streamlines are shown in the figure:

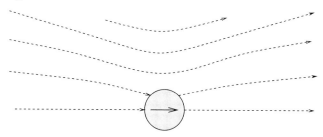

Apparently, the main difference is that inviscid streamlines are loops (compare with the loops made by trajectories as shown in Figure 1.9), while viscous flow is one-way.

In the sphere reference frame, the Stokes stream function is $\psi = -ur^2\sin^2\theta(1/2 - 3R/4r + R^3/4r^3)$ and the streamlines are qualitatively similar to those in the right part of Figure 4.1.

1.7 The equation of motion for the ball on a spring is $m\ddot{x} = -kx$ and the corresponding frequency is $\omega_a = \sqrt{k/m}$. In a fluid,

$$m\ddot{x} = -kx - \tilde{m}\ddot{x}, \qquad (4.6)$$

where $\tilde{m} = \rho V/2$ is the induced mass of a sphere. The frequency of oscillations in an ideal fluid is

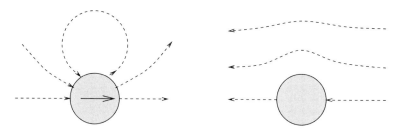

Figure 4.1 Streamlines of the potential flow around a sphere in the reference frame where the fluid is at rest at infinity (left) and in the reference frame moving with the sphere (right).

$$\omega_{a,\text{fluid}} = \omega_a \sqrt{\frac{2\rho_0}{2\rho_0 + \rho}}; \qquad (4.7)$$

here ρ is fluid density and ρ_0 is the ball's density. Mechanical watch runs faster in the air of lower density (as was noticed by I. Rabi on a mountain and presented as a problem to E. Fermi, who solved it immediately).

The equation of motion for the pendulum is $ml\ddot{\theta} = -mg\theta$. In a fluid, it is

$$ml\ddot{\theta} = -mg\theta + \rho V g\theta - \tilde{m}l\ddot{\theta} \qquad (4.8)$$

where $\rho V g\theta$ is the Archimedes force and $-\tilde{m}l\ddot{\theta}$ the inertial force. From $ml\ddot{\theta} = -mg\theta$ we get $\omega_b = \sqrt{g/l}$, while when placed in the fluid we have that the frequency of oscillations is

$$\omega_{b,\text{fluid}} = \omega_b \sqrt{\frac{2(\rho_0 - \rho)}{2\rho_0 + \rho}}. \qquad (4.9)$$

The fluid viscosity would lead to some damping. When the viscosity is small, $v \ll \omega_{a,b}a^2$, then the width of the boundary layer is much less then the size of the body: $v/\omega_{a,b}a \ll a$. We can then consider the boundary layer as locally flat and for a small piece near a flat surface we derive

$$\frac{\partial v_x}{\partial t} = v \frac{\partial^2 v_x}{\partial y^2},$$

$$v_x(y,t) = u \exp\{-[(1+\mathrm{i})y/\delta + \mathrm{i}\omega t]\}, \quad \delta = \sqrt{2\eta/\rho_0 \omega}. \qquad (4.10)$$

Such fluid motion provides for the viscous stress on the body surface

$$\sigma_{yx} = \eta \frac{\partial v_v(0,t)}{\partial y} = (\mathrm{i} - 1)v_x(0,t)\sqrt{\omega\eta\rho/2},$$

which leads to the energy dissipation rate per unit area

$$-\sigma_{yx}v_y = u^2\sqrt{\omega\eta\rho/8}.$$

An estimate of the energy loss can be obtained by multiplying it by the surface area. To get an exact answer for the viscous dissipation by the oscillating sphere, one needs to find the velocity distribution around the surface, see e.g. Chapter 24 of [16].

1.8 **Dimensional analysis and simple estimates.** In the expression $T \propto E^\alpha p^\beta \rho^\gamma$, the three unknowns α, β, γ can be found from considering three dimensionalities (grams, metres and seconds), which gives

$$T \propto E^{1/3} p^{-5/6} \rho^{1/2} \, .$$

Analogously, $T \propto a p^{-1/2} \rho^{1/2}$. Note that here $c \propto \sqrt{p/\rho}$ is the sound velocity so that the period is a/c. The energy is pressure times volume: $E = 4\pi a^3 p/3$. This method is used to measure the energy of the explosions underwater: wait until the bubble is formed and then relate the bubble size, obtained by measuring bubble oscillations, to the energy of the explosion.

Sketch of a theory. The radius of the bubble varies as: $r_0 = a + b\exp(-i\omega t)$, where a is the initial radius and $b \ll a$ is a small amplitude of oscillations with the period $T = 2\pi/\omega$. The induced fluid flow is radial, if we neglect gravity, $v = v_r(r,t)$. Incompressibility requires $v(r,t) = A\exp(-i\omega t)/r^2$. On the surface of the bubble, $dr_0/dt = v(r,t)$, which gives $A = -iba^2\omega$. So the velocity is

$$v(r,t) = -ib(a/r)^2 \omega \exp(-i\omega t). \qquad (4.11)$$

Note that $(\mathbf{v} \cdot \nabla)\mathbf{v} \simeq b^2 \omega^2/a \ll \partial_t \mathbf{v} \simeq b\omega^2$, since it corresponds to the assumption $b \ll a$. Now we use the linearized Navier–Stokes equations and spherical symmetry and get:

$$p_{\text{water}} = p_{\text{static}} - \rho\omega^2 b \left(\frac{a^2}{r} \right) e^{-i\omega t}, \qquad (4.12)$$

where p_{static} is the static pressure of water – unperturbed by oscillations. The period of oscillations has been estimated as the radius of the bubble divided by the speed of sound in the water, which is supported by (4.16) below. The speed of sound in the air bubble is much larger than in the water because the pressure is the same while the air density is much lower. Therefore, the pressure inside the bubble equilibrates fast and can be considered uniform. On the other hand, we assume that the period of oscillations is much smaller than the heat exchange time, which is the radius squared divided by the thermal diffusivity. In distinction from the viscosity (see Section 1.4.3), the diffusivity in a liquid requires the transfer of molecules and can be estimated as the speed of sound times mean free path. Therefore, the heat exchange is negligible when the radius is much larger than the mean free path. That allows us to use the adiabatic law, $p_{\text{bubble}} r_0^{3\gamma} = p_{\text{static}} a^{3\gamma}$, which gives:

$$p_{\text{bubble}} = p_{\text{static}} \left(1 - 3\gamma(b/a)e^{-i\omega t} \right). \qquad (4.13)$$

Now use the boundary condition for the bubble–water interface at $r = a$

$$-p_{\text{bubble}}\delta_{ik} = -p_{\text{water}}\delta_{ik} + \eta\left(\partial_k v_i + \partial_i v_k\right), \qquad (4.14)$$

where η is the water dynamic viscosity. The component σ_{rr} gives $\rho(a\omega)^2 + 4i\eta - 3\gamma p = 0$ with the solution

$$\omega = \left(-\frac{2\eta i}{a^2 \rho} \pm \sqrt{\frac{3\gamma p}{\rho a^2} - \frac{4\eta^2}{a^4 \rho^2}} \right). \tag{4.15}$$

For large viscosity, it describes aperiodic decay; for $\eta^2 < 3\gamma p \rho a^2 / 4$ the frequency of oscillations is:

$$\omega = 2\pi / T(a, p, \rho) = \sqrt{\frac{3\gamma p}{\rho a^2} - \frac{4\eta^2}{a^4 \rho^2}}. \tag{4.16}$$

Viscosity slows down oscillations i.e. increases the period, which may be relevant for small bubbles, see [31] for more details.

1.9 The Navier–Stokes equation for the vorticity in an incompressible fluid,

$$\partial_t \omega + (\mathbf{v} \cdot \nabla)\omega - (\omega \cdot \nabla)\mathbf{v} = \nu \Delta \omega$$

in the cylindrically symmetric case is reduced to the diffusion equation,

$$\partial_t \omega = \nu \Delta \omega,$$

since $(\mathbf{v} \cdot \nabla)\omega = (\omega \cdot \nabla)\mathbf{v} = 0$. The diffusion equation in the cylindrical reference frame has the form

$$\partial_t \omega = \nu r^{-1} \partial_r r \partial_r \omega.$$

With the initial condition as a delta function on a line, the equation has the spreading-vortex solution

$$\omega(r, t) = \frac{\Gamma}{4\pi\nu t} \exp\left(-\frac{r^2}{4\nu t} \right),$$

which conserves the total vorticity:

$$\Omega(t) = 2\pi \int_0^\infty \omega(r, t) r dr = \frac{\Gamma}{2\nu t} \int_0^\infty \exp\left(-\frac{r^2}{4\nu t} \right) r dr = \Gamma.$$

Generally, for any two-dimensional incompressible flow, the Navier–Stokes equation takes the form $\partial_t \omega + (\mathbf{v} \cdot \nabla)\omega = \nu \Delta \omega$, which conserves the vorticity integral, which is also the vorticity flux, as long as it is finite.

1.10 The shape of the swimmer is characterized by the angles between the arms and the middle link. Therefore, the configuration space is two-dimensional. Our swimmer goes around a loop in this space with the displacement proportional to the loop area, which is $\Delta\theta_1 \Delta\theta_2 = \theta^2$. The transformation $\theta_1 \rightarrow -\theta_1$, $\theta_2 \rightarrow -\theta_2$, $y \rightarrow -y$ produces the same loop; since the displacement does not change under $\theta_1 \to -\theta_1$, $\theta_2 \to -\theta_2$, then the y-displacement must be zero. Alternatively, one may argue that this sequence of configurations is invariant with respect to the transformation $t \to -t$, $x \to -x$, $y \to y$ so the displacement must be along x. Apparently, arm movements do not commute. For small angles, it is easier to move along x when the non-moving arm is aligned with the body (i.e. either θ_1 or θ_2 is zero). Therefore, the swimmer shifts to the left during $1 \to 2$ and $4 \to 5$ less than it shifts to the right during $2 \to 3$ and $3 \to 4$, at least when $\theta \ll 1$. The total displacement is to the right or generally in the direction of the arm that moved first. Further reading: section 7.5 of [2, 9, 26].

Anchoring the swimmer, one gets a pump. The geometrical nature of swimming and pumping by micro-organisms makes them a subject of a non-Abelian gauge field theory with rich connections to many other phenomena, see [33].

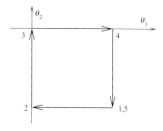

1.11 **Simple estimate.** The angular velocity is $\Omega = 20\pi$ rad/sec. The lift force can be estimated as $\rho v_0 \Omega R R^2 \simeq 2.64$ N. The deflection can be roughly estimated by neglecting the ball deceleration and estimating the time of flight as $T \simeq L/v_0 \simeq 1$ s. Further, neglecting the drag in the perpendicular direction we estimate that the acceleration $\rho v_0 \Omega R^3/m \simeq 5.9$ m·s^{-2} causes the deflection

$$y(T) = \frac{\rho v_0 \Omega R^3 T^2}{2m} \simeq \frac{\rho \Omega R^3 L^2}{2mv_0} = L\frac{\Omega R}{v_0}\frac{\rho R^2 L}{2m} \simeq 2.9 \text{ m} . \quad (4.17)$$

Sketch of a theory. It is straightforward to account for the drag force, $C\rho v^2 \pi R^2/2$, which leads to the logarithmic law of displacement:

$$\dot{v} = -\frac{v^2}{L_0} , \quad v(t) = \frac{v_0}{1 + v_0 t/L_0} ,$$
$$x(t) = L_0 \ln(1 + v_0 t/L_0) . \quad (4.18)$$

Here $L_0 = 2m/C\rho\pi R^2 \simeq 100$ m with $C \simeq 0.25$ for $Re = v_0 R/\nu \simeq 2 \cdot 10^5$. The meaning of the parameter L_0 is that this is the distance at which drag substantially affects the speed; unsurprisingly, it corresponds to the mass of the air displaced, $\rho \pi R^2 L$, being of the order of the ball's mass. We get the travel time T from (4.18): $v_0 T / L_0 = \exp(L/L_0) - 1 > L/L_0$. We can now account for the time dependence $v(t)$ of the deflection. Assuming that the deflection in the y-direction is small compared with the path travelled in the x-direction, we get

$$\frac{\mathrm{d}^2 y(t)}{\mathrm{d}t^2} = \frac{\rho \Omega R^3 v(t)}{m} = \frac{\rho \Omega R^3 v_0}{m(1 + v_0 t / L_0)},$$

$$y(0) = \dot{y}(0) = 0,$$

$$y(t) = L_0 \frac{\Omega R}{v_0} \frac{2}{\pi C} [(1 + v_0 t / L_0) \ln(1 + v_0 t / L_0) - v_0 t / L_0]. \quad (4.19)$$

It turns into (4.17) in the limit $L \ll L_0$ (works well for a penalty kick). For longer L, one also needs to account for the drag in the y-direction, which leads to saturation of $\dot{y}$ at the value $\sim \sqrt{\Omega R v}$. Still, such detailed consideration does not make much sense because we took a very crude estimate of the lift force and neglected vertical displacement $gT^2/2$, which is comparable with the deflection.

   Remark. Great soccer players are also able to utilize the drag crisis, which is a sharp increase of the drag coefficient C from 0.15 to 0.5 when Re decreases from $2.5 \cdot 10^5$ to $1.5 \cdot 10^5$ (the ball's velocity drops from 37.5 m s^{-1} to 22.5 m s^{-1}). As a result, the ball sharply decelerates some way into its path and the Magnus force comes even more into effect. The phenomenon of drag crisis is also used for making a long shot over a goalkeeper who has come out too far from the goal; in this case, the ball smoothly rises up and then falls down steeply (see figure). A top spin adds Magnus force, which enhances this effect. Similarly, a tennis ball with a proper topspin dips sharply after passing over the net.

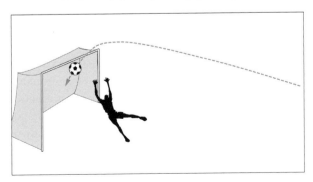

1.12 The force on the keel exists only if the board does not move in exactly the direction in which it is pointed. The direction of the force acting on the keel can be understood by considering the deflection of water in the reference frame of the keel – water comes from the direction of motion and leaves along the keel. The direction of the force acting on the keel is opposite to the direction of deflection of water by the keel. That force, $\mathbf{F}_{keel}$, acting mainly to the side (left in the figure), must be counteracted by the force of the wind acting on the sail. The wind leaves along the sail and its deflection determines $\mathbf{F}_{sail}$. For a board in a steady motion, the vector $\mathbf{F}_{keel} + \mathbf{F}_{sail}$ points in the direction of motion and it is balanced by the drag force. Decreasing the drag, one can in principle move faster than the wind since $\mathbf{F}_{sail}$ does not depend on the board speed (as long as one keeps the wind perpendicular *in the reference frame of the board*). On the contrary, when the board moves downwind, it cannot move faster than the wind.

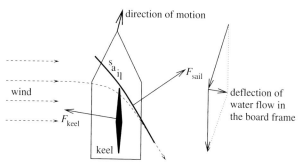

   For details, see [3].

1.13 (i) **Simple answer**. If the droplet is a solid sphere, one uses the Stokes force and gets the steady fall velocity from the force balance

$$6\pi R\eta_a u = mg, \quad u = \frac{2\rho_w g R^2}{9\eta_a} = g\tau \simeq 1.21\,\mathrm{cm\,s^{-1}}. \quad (4.20)$$

Stokes himself used his formula to explain (in a passing remark) suspension of clouds: small droplets fall very slow since the weight decreases with the radius much faster than the resistance.

   Applicability limits and correction. The Reynolds number is $Re \simeq 0.008$, which justifies our approach and guarantees that we can neglect finite-Re corrections. Note that $Re \propto vR \propto R^3$, so that $Re \simeq 1$ already for $R = 0.05\,\mathrm{mm}$. Sphericity is maintained by surface tension: the relevant parameter is the ratio of the viscous stress $\eta_w u/R$ to the surface tension stress α/R; this ratio is

$\eta_w u/\alpha \simeq 0.00017$ for $\alpha = 70\,\mathrm{g\,s^{-2}}$. Another unaccounted phenomenon is internal circulation in a liquid droplet. The viscous stress tensor $\sigma_{r\theta}$ must be continuous through the droplet surface, so that the velocity inside can be estimated as the velocity outside times the small factor $\eta_a/\eta_w \simeq 0.018 \ll 1$, which is expected to give a 2% correction to the force and to the fall velocity. Let us calculate this. The equation for the motion of the fluid inside is the same as outside. The regular solution at infinity is (1.58), i.e.

$$\mathbf{v_a} = \mathbf{u} - a\frac{\mathbf{u} + \hat{\mathbf{n}}(\mathbf{u}\cdot\hat{\mathbf{n}})}{r} + b\frac{3\hat{\mathbf{n}}(\mathbf{u}\cdot\hat{\mathbf{n}} - \mathbf{u})}{r^3},$$

while the regular solution at zero is $f = Ar^2/4 + Br^4/8$, which gives

$$\mathbf{v_w} = -A\mathbf{u} + Br^2(\hat{\mathbf{n}}(\mathbf{u}\cdot\hat{\mathbf{n}})) - 2\mathbf{u}.$$

Four boundary conditions on the surface (zero normal velocities and continuous tangential velocities and stresses) fix the four constants A, B, a, b and give the drag force

$$F = 8\pi a\eta u = 2\pi u\eta_a R\frac{2\eta_a + 3\eta_w}{\eta_a + \eta_w}, \qquad (4.21)$$

which leads to

$$u = \frac{2\rho R^2 g}{3\eta_a}\left(\frac{3\eta_a + 3\eta_w}{2\eta_a + 3\eta_w}\right) \simeq \frac{2\rho R^2 g}{9\eta_a}\left(1 + \frac{1}{3}\frac{\eta_a}{\eta_w}\right) \simeq 1.22\,\mathrm{cm\,s^{-1}}.$$

The inside circulation acts as a lubricant, decreasing drag and increasing fall velocity. In reality, however, water droplets often fall as solid spheres because of a dense "coat" of dust particles accumulated on the surface.

(ii) **Basic solution.** Denote the droplet radius r and its velocity v. We need to write conservation of mass $\dot{r} = Av$ and the equation of motion, $dr^3v/dt = gr^3 - Bvr$, where we assumed a low Reynolds number and used the Stokes expression for the friction force. Here A, B are some constants. One can exclude v from here but the resulting second-order differential equation is complicated. To simplify, we assume that the motion is quasi-steady so that gravity and friction almost balance each other out. This requires $\dot{v} \ll g$ and gives $v \approx gr^2/B$. Substituting that into the mass conservation gives $dr/dt = Agr^2/B$. The solution of this equation gives an explosive growth of the particle radius and velocity: $r(t) = r_0/(1 - r_0Agt/B)$ and $v = v_0/(1 - r_0Agt/B)^2$. This solution is true as long as $\dot{v} \ll g$ and $Re = vr/\nu \ll 1$.

Detailed solution. Denote ρ_w, ρ, ρ_v, respectively, the densities of the liquid water, air and water vapour. Assume $\rho_w \gg \rho \gg \rho_v$. Mass conservation gives $dm = \rho_v \pi r^2 v dt = \rho_w 4\pi r^2 dr$ so that $dr/dt = v\rho_v/4\rho_w$. Initially, we may consider low-Re motion so that the equation of motion is as follows: $dr^3 v/dt = gr^3 - 9v\rho vr/2\rho_0$. Quasistatic approximation is $v \approx 2gr^2 \rho_w/9v\rho$ according to (4.20), which gives the equation $dr/dt = gr^2 \rho_v/18v\rho$ independent of ρ_w. The solution of this equation gives the quantitative description of the explosive growth of the particle radius and velocity:

$$r(t) = r_0 \left(1 - \frac{\rho_v}{\rho} \frac{r_0 g t}{18v} \right)^{-1}, \quad v(t) = \frac{\rho_w}{\rho} \frac{2gr_0^2}{9v} \left(1 - \frac{\rho_v}{\rho} \frac{r_0 g t}{18v} \right)^{-2}.$$

This solution is true as long as $\dot{v} = 4gr\dot{r}\rho_0/9v\rho = 2g^2 r^3 \rho_v \rho_0/81v^2\rho^2 \ll g$. Also, when $r(t)v(t) \simeq v$ the regime changes so that $mg = C\rho \pi r^2 v^2$, $v \propto \sqrt{r}$ and $r \propto t^2$, $v \propto t$.

1.14 The pressure is constant along the free jet boundaries and so the velocity is constant as well. Therefore, the asymptotic velocities in the outgoing jets are the same as in the incoming jets in the reference frame A, where the meeting point is a stagnation point. Conservation of mass and horizontal momentum are $2h = h_r + h_l$ and $2h\cos\theta_0 = h_l - h_r$, where h is the width of the incoming jet. That gives for the left and right jets respectively

$$h_l = h(1 + \cos\theta_0), \quad h_r = h(1 - \cos\theta_0).$$

Therefore, a fraction $(1 - \cos\theta_0)/2$ of the metal cone is injected into the forward jet.

For the cumulative jet, detonation of explosives actually determines the compression speed of the cone, that is the velocity U normal to the cone in its reference frame. All jets have velocity V on their surface. The cone motion is normal to itself in the reference frame B moving to the left with the speed $V/\cos\theta$, where the normal velocity is $V\tan\theta = U$. In this frame, the cumulative jet moves with the speed $v = V(1 + 1/\cos\theta) = U(1 + \cos\theta)/\sin\theta$. Apparently, the smaller the angle, the larger the jet speed. However, the momentum of the jet is proportional to $vh_r = hU\sin\theta$, that is increases with the angle. In most shells $U \simeq 2$ km/sec, while the jet speed can be tens of km/sec; upon hitting armor it produces pressures $\rho v^2/2$ in excess of a million bars. The pressure is so high that inertia dominates elasticity, and the flow can indeed be considered incompressible (until the material vaporizes).

One can describe the whole flow field in terms of the complex velocity **v**. In the reference frame A, the velocity **v** changes inside a circle whose radius is the velocity at infinity, V (see e.g. Chapter XI of [20]). On the circle, the stream function is piecewise constant with the jumps equal to the jet fluxes:

$$\psi = 0 \quad \text{for} \quad 0 \le \theta \le \theta_0, \qquad \psi = -hV \quad \text{for} \quad \theta_0 \le \theta \le \pi,$$

$$\psi = (h_l - h)V \quad \text{for} \quad \pi \le \theta \le 2\pi - \theta_0, \text{ etc.}$$

We can now find the complex potential everywhere in the circle by using the Schwartz formula:

$$w(\mathbf{v}) = \frac{i}{2\pi} \int_0^{2\pi} \psi(\theta) \frac{V \exp(i\theta) + \mathbf{v}}{V \exp(i\theta) - \mathbf{v}} d\theta$$

$$= \frac{V}{\pi} \left\{ h_r \ln \left(1 - \frac{\mathbf{v}}{V} \right) + h_l \ln \left(1 + \frac{\mathbf{v}}{V} \right) \right.$$

$$\left. - h \ln \left[\left(1 - \frac{\mathbf{v}}{V} e^{i\theta_0} \right) \left(1 - \frac{\mathbf{v}}{V} e^{-i\theta_0} \right) \right] \right\}.$$

To relate the space coordinate z and the velocity **v** we use $\mathbf{v} = -dw/dz$ so that one needs to differentiate $w(\mathbf{v})$ and then integrate once the relation $dz/d\mathbf{v} = -\mathbf{v}^{-1} dw/d\mathbf{v}$ using $z = 0$ at $\mathbf{v} = 0$:

$$\frac{\pi z}{h} = (1 - \cos\theta_0) \ln \left(1 - \frac{\mathbf{v}}{V} \right) - (1 + \cos\theta_0) \ln \left(1 + \frac{\mathbf{v}}{V} \right)$$

$$+ e^{i\theta_0} \ln \left(1 - \frac{\mathbf{v}}{V} e^{i\theta_0} \right) + e^{-i\theta_0} \ln \left(1 - \frac{\mathbf{v}}{V} e^{-i\theta_0} \right).$$

1.15 Viscous friction between fluid layers leads to momentum exchange conserving the total momentum. Fluid rotation corresponds to only axial velocity v_θ being nonzero. The axial component of the Navier–Stokes equation in the cylindrical coordinates has the form

$$\frac{\partial v_\theta}{\partial t} = \nu \left(\Delta v_\theta - \frac{v_\theta}{r^2} \right) = \frac{\nu}{r^2} \frac{\partial}{\partial r} r^3 \frac{\partial}{\partial r} \frac{v_\theta}{r}. \qquad (4.22)$$

We met the curl of this equation for $\omega = r^{-1} \partial_r r v_\theta$ in the Problem 1.9. The equation (4.22) conserves the angular momentum $\int_0^\infty v_\theta r^2 dr$. It has two steady solutions: i) one with a singularity at zero radius, $v_\theta \propto 1/r$, which has vorticity as a delta-functions; ii) another with a singularity at infinity, $v_\theta \propto r$, which corresponds to a constant vorticity and a solid-body rotation. If initially we have an angular momentum localized in some region, viscosity will make it spread. Only if a wall at rest is present then viscous friction will make the momentum absorbed.

The fact that viscosity redistributes momentum may lead to surprising phenomena. For example, if one opens a small hole at the center of the bottom of a rotating vessel, it starts to rotate faster. Indeed, if an ideal fluid moves toward the hole, it accelerates its rotation because the angular momentum is conserved for any particle (which we all observed draining a bathtub). Viscous friction tends to equilibrate the angular momentum across the radius, thus decelerating the inner layers while accelerating the outer layers and the vessel itself. Believe it or not, total angular momentum per unit mass increases with time since the fluid that leaves the vessel has less momentum density than the remaining fluid.

1.16 Since the velocity is potential, $v_i \propto x_i r^{-d} \propto \partial_i r^{2-d}$, then the vorticity is zero. Velocity derivatives are nonzero and so are the components of the viscous stress tensor: $\sigma_{ik} = \eta \left[\partial_k v_i + \partial_i v_k \propto (\delta_{ik} - dx_i x_k / r^2) r^{-d} \right]$. In particular, for $d = 3$ one has $\sigma_{\phi\phi} = \sigma_{\theta\theta} = -\sigma_{rr}/2 \propto r^{-3}$. The net viscous force acting on any fluid element is the divergence of the stress tensor, that is the Laplacian of the velocity, which is zero as for any potential incompressible flow:

$$\Delta v_i = \sum_k \partial_k^2 v_i \propto [d(d+2) - d^2] x_i x_l x_m \delta_{lk} \delta_{mk} r^{-d-4} - 2dx_j \delta_{ik} \delta_{jk} r^{-d-2} = 0.$$

That can be seen in spherical coordinates as well, if to remember that the Laplacian of the vector in curvilinear coordinates includes the derivative of the unit vector:

$$(\Delta \mathbf{v})_r = \Delta v_r - (d-1) v_r / r^2 = r^{1-d} \partial_r r^{d-1} \partial_r v_r - (d-1) v_r / r^2 = 0.$$

Since the net viscous force is zero everywhere, then any fluid element moves without acceleration, yet is deformed by stresses acting on it: compressed along the radial direction and expanded transversally. That deformation is accompanied by energy dissipation whose integral is determined by the size of the source r_0:

$$\eta \int (\partial_k v_i + \partial_i v_k)^2 d^d r \propto \eta r_0^{-d}.$$

To compensate for the energy loss, some force must do the work. However, the pressure stays constant inside the low-Re viscous potential flow: $\nabla p = \eta \Delta \mathbf{v} = 0$. That means the pressure gradient which does the work is inside the source or concentrated at its surface (within the layer comparable to the mean free path). The pressure times total current must be equal to the dissipation, which gives the pressure $p_0 \propto \eta r_0^{-d}$. One can consider the limit of a point source, then $\nabla p = \eta \Delta \mathbf{v} \propto \nabla \delta(\mathbf{r})$.

Let us briefly discuss the validity of low-Re approximation and the role of inertial effects. Formally, one can estimate $Re(r) = vr/v \propto r^{2-d}$ so that the low-Re approximation is either valid uniformly (at $d = 2$) or improves away from the source (at $d > 2$). However, since the viscous force is identically zero, the inertial term is the only one that contributes the pressure in the bulk: $p(r) - p(\infty) = -v^2/2 \propto r^{2-2d}$. It decreases toward the source, but its relative contribution to p_0 is small as long as $Re(r_0)$ is small.

An even simpler example of a force-free flow in the bulk sustained by the work done at the boundaries is a constant-shear flow that is a linear velocity profile. Viscous force and pressure gradient are identically zero yet the stress is not. Every fluid element deforms as described in Section 1.2.2. Such flow needs a moving plane boundary that does the work to sustain the flow. Another example is the potential flow of surface wave, where, according to Stokes, the bulk dissipation is equal to the work by the surface stress, as described in Section 3.1.2.

1.17 For a steady flow, the pressure gradient must be equal to the momentum flux to the walls, which absorb momentum. The momentum flux can be estimated as the momentum divided by the time it takes for momentum to get to the wall. So the momentum and the mass flow rate at fixed pressure gradient are proportional to the momentum transfer time. At low density when the mean free path l is much larger than w, the time is the pipe radius w divided by the thermal velocity v_T and is independent of the density (except for a perfectly straight pipe, see the remark below). At high density, when the mean free path l is much smaller than w, the momentum must diffuse to the wall through multiple collisions and the time is viscous: $w^2/v \simeq w^2/v_T l$. The viscous time is by the factor w/l larger than the ballistic time w/v_T, so that the flow rate is larger for a viscous Poiseuille flow than for a Knudsen flow of a rarefied gas. Mean-free path is inversely proportional to the density so that the viscous time and the flow rate linearly increase with the density. One can look at it also from a spatial perspective. Indeed, the Knudsen flow profile is uniform across the pipe, while the Poiseuille flow profile is convex with the velocity having maximum at the center and zero at the wall. In other words, interaction between particles makes them flow staying away from the walls, where the momentum loss occurs, thus decreasing resistance. The same effect occurs for electrons whose interaction can help them avoid scatterers and boundaries and increase conductivity above the ballistic value.

Further increasing gas density and decreasing viscosity, one comes eventually into the regime of large Reynolds numbers. Transition to turbulence strongly enhances friction factor and diminishes the flow rate.

Remark. For a perfectly straight pipe, the resistance at $l \gg w$ depends weakly (logarithmically) on the mean free path due to particles flying at small angles θ to the walls. Indeed, the flight time for a given angle is $w/v_T \theta$ as long as $\theta > w/l$. The time averaged over the angle diverges logarithmically when $l \to \infty$:

$$\frac{w}{v_T} \int_{w/l}^{\pi/2} \frac{d\theta}{\theta} \approx \frac{w}{v_T} \ln \frac{l}{w}.$$

As a result, the resistance is maximal and the flow rate is minimal for $l \simeq w$ (Knudsen paradox).

1.18 **Dimensions and limits.** With five parameters, E, ρ, c, h, r_0, and three independent dimensions, cm, s and g, one can combine two different dimensionless parameters, according to the π-theorem of Section 1.4.4. Then there is no hope for a simple answer in a general case. Since we are interested in the limit $r_0/h \to 0$, consider first the simplest case of a spherical explosion in an unbounded medium ($h \to \infty$) to establish how the dimensional parameters E, ρ, c, r_0 combine into a dimensionless parameter. Such spherical flow was described in Exercise 1.16: $v = Ar^{-2}$. Calculating its energy, $E = \rho \int v^2 \, d^3r/2 = A^2 2\pi\rho/r_0$, we find

$$A = \sqrt{Er_0/2\pi\rho}. \tag{4.23}$$

That tells, in particular, that the flows are the same for the same values of Er_0, that is to use less explosives one needs larger void. Returning to our problem in the limit of point charge, $r_0/h \to 0$, we have three dimensional parameters, Er_0/ρ, c, h, and two independent dimensions, cm, s. That means that now there is only one dimensionless parameter: $\xi = Er_0/2\pi\rho h^4 c^2$. Here $Er_0/2\pi\rho h^4 \equiv u^2$ can be interpreted as the typical squared velocity, created by the explosion at the distance h, so that $\xi = (u/c)^2$. The crater radius is then $R = hf(\xi)$. We expect no crater for $\xi \ll 1$ when the critical velocity at the surface is not reached, which happens for very weak or very deep explosions. In the opposite limit of strong and shallow explosions, we expect the vertical velocity to decay with the distance R along the surface as h/R^3. Indeed, the velocity of incompressible flow decays as R^{-2}, and for vertical velocity at $R \gg h$ we expect extra small factor h/R. The ratio of the surface velocities in the center and at the edge of a crater is then $u/c \propto R^3/h^3$ so that $R/h \propto \xi^{1/6}$.

Sketch of a theory. The velocity potential of a point explosion is the same as the electric potential of a point charge. Recall that after a short pulse of pressure, the potential is equal to $-\int dt\, p/\rho$, see (1.45). The pressure is constant on the soil surface and so must be the potential. That means that the horizontal velocity must be zero on the soil boundary – if you watch war movies, notice how an underwater explosion ejects water as a vertical column. In electrostatics, equipotential boundary is metallic. The potential in space is that of a point source (charge) plus its mirror image of a point sink, written in cylindrical coordinates as follows:

$$\phi(r,z) = A\left[r^2 + (z-h)^2\right]^{-1/2} - A\left[r^2 + (z+h)^2\right]^{-1/2}. \tag{4.24}$$

The constant must be found from evaluating the energy of the flow and it is the same (4.23). Indeed, the boundary does not change the energy, just redirects it; alternatively, one can argue that this is the answer in the limit $h \to \infty$ and it must be independent of h because there is no other length parameter. For those who prefer a direct computation, here it is:

$$E = \frac{\rho}{2}\int v^2 d^3r = \frac{\rho}{2}\int \operatorname{div}(\phi\nabla\phi)\,d^3r = \frac{\rho}{2}\oint \phi\nabla\phi\cdot d\mathbf{f}.$$

The last integral is over the boundary which consists of two parts: plane $z=0$ where $\phi=0$ and sphere with the radius r_0 and center at $(0,0,-h)$. Only the first term in (4.24) contributes the integral over the spherical surface.

The vertical velocity at $z=0$ is $v(r,0) = 2Ah/(r^2+h^2)^{3/2}$. The condition $v(R,0)=c$ gives the crater radius

$$R^2/h^2 = (4\xi)^{1/3} - 1 = \left(\frac{2Er_0}{\pi\rho c^2 h^4}\right)^{1/3} - 1. \tag{4.25}$$

We see that indeed there is a threshold and that the minimal energy for crater creation grows as the fourth power of the burial depth h. One can also find the crater depth d from the condition $v(0,d)=c$ which gives $d = 2^{1/3}R$ for shallow explosions. How well (4.25) describes craters in the Nevada Test Site one can see in M. Feit et al, 2001, SPIE **4347**, 316. That description also applies to the damage done by powerful radiation absorbed by small imperfections in optical materials – apparently, most damage is done by near-surface absorption.

Remark. Another beautiful application of the theory of incompressible potential flows is the engineering feat of directed explosion. Let us ask: how to distribute explosives over the given convex surface $z = S(x,y)$ to impact a uniform velocity $\mathbf{U} = (U,V,W)$ onto a material inside this surface to move it as one piece? On the surface, the density of the explosive, $\rho\left(x,y,S(x,y)\right)$, is proportional to the pressure, which in its own turn is proportional to the velocity potential. Inside the moving fluid the velocity is constant so that the potential is linear in coordinates and $\rho\left(x,y,S(x,y)\right) \propto Ux+Vy+WS(x,y)+$ const. That was used by Lavrentiev to move a large amount of soil and block mountain torrents in Medeo canyon, see Barenblatt, G. I. 2014. *Flow, Deformation and Fracture* (Cambridge: Cambridge University Press).

4.2 Chapter 2

2.1 We have seen in Section 1.2.2 that in a locally smooth flow, fluid elements either stretch or contract exponentially in a strain-dominated flow or rotate in a vorticity-dominated flow. This is true also for flows in phase space, discussed at the beginning of Section 2.2.

(i) Since $x(t)=x_0\exp(\lambda t)$ and $y(t)=y_0\exp(-\lambda t)=x_0y_0/x(t)$ then every streamline (and trajectory) is a hyperbola. A vector initially forming an angle φ with the x-axis will be stretched after time T if $\cos\varphi \geq [1+\exp(2\lambda T)]^{-1/2}$, i.e. the fraction of stretched directions is greater than half. This means, in particular, that if one randomly changes direction after some time, the net effect is stretching.

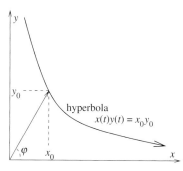

(ii) The eigenvectors of this problem evolve according to $\exp(\pm i\Omega t)$, where

$$\Omega^2 = 4\omega^2 - \lambda^2. \qquad (4.26)$$

We see that fluid rotates inside vorticity-dominated (elliptic) regions and is monotonically deformed inside strain-dominated (hyperbolic) regions. The marginal case is a shear flow (see Figure 1.6) where $\lambda=2\omega$ and the distances grow linearly with time.

In a random flow (in either real space or phase space), a fluid element visits many different elliptic and hyperbolic regions along its way. After a long random sequence of deformations and rotations, we find it stretched into a thin strip. Of course, this is a statistical statement, meaning that the probability of finding a ball turning into an exponentially stretching ellipse goes to unity as time increases. This is because the larger the ellipse eccentricity the more likely stretching happens along the long axis. After the strip length reaches the scale of the velocity change (when one already cannot approximate the velocity by a linear profile), the strip starts to fold continuing locally the exponential stretching. Eventually, one can find points from the initial ball everywhere, which means that the flow is mixing.

2.2 **Dimensional reasoning.** With six parameters, $g, \beta, \Theta, h, v, \chi$ and three independent dimensions, cm, s and kelvin, one can combine three different dimensionless parameters, according to the π-theorem of Section 1.4.4. This is too many parameters for any meaningful study.

Basic physical reasoning suggests that the first three parameters can combine only as a product, $g\beta\Theta$, which is a buoyancy force per unit mass (the density cannot enter because there is no other parameter having mass units). We now have four parameters, $\beta g\Theta, h, v, \chi$ and two independent dimensions, cm, s, so we can make two dimensionless parameters. The first one characterizes the medium and is called the Prandtl number:

$$Pr = v/\chi \ . \tag{4.27}$$

The same molecular motion is responsible for the diffusion of momentum by viscosity and the diffusion of heat by thermal conductivity. Nevertheless, the Prandtl number varies greatly from substance to substance. For gases, one can estimate χ as the thermal velocity times the mean free path, exactly like viscosity in Section 1.4.3, so that the Prandtl number is always of order unity. For liquids, Pr varies from 0.044 for mercury through 6.75 for water to 7250 for glycerol.

The second parameter can be constructed in infinitely many ways as it can contain an arbitrary function of the first parameter. One may settle on any such parameter, claiming that it is a good control parameter for a given medium (for fixed Pr). However, one can do better than that and find the control parameter that is the same for all media (i.e. all Pr). The physical reasoning helps one to choose the right parameter. It is clear that convection can occur when the buoyancy force, $\beta g\Theta$, is larger than the friction force, vv/h^2. It may seem that taking velocity v small

enough, one can always satisfy that criterion. However, one must not
forget that as the hotter fluid rises it loses heat by thermal conduction
and becomes more dense. Our estimate of the buoyancy force is valid as
long as the conduction time, h^2/χ, exceeds the convection time, h/v, so
that the minimal velocity is $v \simeq \chi/h$. Substituting that velocity into the
friction force, we obtain the correct dimensionless parameter as the
force ratio, which is called the Rayleigh number:

$$Ra = \frac{g\beta\Theta h^3}{\nu\chi} \ . \tag{4.28}$$

Sketch of a theory. The temperature T satisfies the linear
convection–conduction equation

$$\frac{\partial T}{\partial t} + (\mathbf{v}\cdot\nabla)T = \chi\Delta T \ . \tag{4.29}$$

For the perturbation $\tau = (T - T_0)/T_0$ relative to the steady profile
$T_0(z) = -\Theta z/h$, we obtain

$$\frac{\partial\tau}{\partial t} - v_z\Theta/h = \chi\Delta\tau \ . \tag{4.30}$$

Since the velocity is itself a perturbation, it satisfies the
incompressibility condition, $\nabla\cdot\mathbf{v} = 0$, and the linearized Navier–Stokes
equation with the buoyancy force:

$$\frac{\partial\mathbf{v}}{\partial t} = -\nabla W + \nu\Delta\mathbf{v} - \beta\tau\mathbf{g} \ , \tag{4.31}$$

where W is the enthalpy perturbation. Of course, the properties of the
convection above the threshold depend on both parameters, Ra and Pr,
so that one cannot eliminate one of them from the system of equations.
If, however, one considers the convection threshold where
$\partial\mathbf{v}/\partial t = \partial\tau/\partial t = 0$, then one can choose the dimensionless variables
$\mathbf{u} = \mathbf{v}h/\chi$ and $w = Wh^2/\nu\chi$ such that the system contains only Ra:

$$-u_z = \Delta\tau \ , \quad \nabla\cdot\mathbf{v} = 0, \quad \frac{\partial w}{\partial z} = \Delta u_z + \tau Ra, \quad \frac{\partial w}{\partial x} = \Delta u_x \ . \tag{4.32}$$

Solving this with proper boundary conditions, for eigenmodes built out
of $\sin(kx), \cos(kx)$ and $\sinh(qz), \cosh(qz)$ (which describe rectangular
cells or rolls), one obtains Ra_{cr} as the lowest eigenvalue, see e.g. [16],
Chapter 57. Near Ra_{cr} the unstable modes has $k \simeq q \simeq 1/h$.

The sufficient condition for convection onset, $Ra > Ra_{\mathrm{cr}}$, is
formulated in terms of the control parameter Ra, which is global that is

a characteristic of the whole system. Note the difference from the local necessary condition (1.9), as found in Section 1.1.3.

2.3 The velocity must be a solution of the Navier–Stokes equation with a constant pressure gradient:

$$\Delta_r v = \frac{1}{r}\frac{d}{dr}r\frac{dv}{dr} = \text{const.}$$

It also must turn into zero at $r = a$:

$$v(r) = 2\bar{v}\left(1 - \frac{r^2}{a^2}\right), \quad \bar{v} = \frac{2\pi}{\pi a^2}\int_0^a v(r)rdr\,. \tag{4.33}$$

The advection–diffusion equation takes the form:

$$\frac{\partial\theta}{\partial t} + v\frac{\partial\theta}{\partial x} = \kappa\Delta\theta\,. \tag{4.34}$$

Averaging it over the cross-section we obtain

$$\frac{\partial\bar{\theta}}{\partial t} + \bar{v}\frac{\partial\bar{\theta}}{\partial x} + \frac{\partial\overline{v\theta'}}{\partial x} = \kappa\frac{\partial^2\bar{\theta}}{\partial x^2}\,. \tag{4.35}$$

Here we assumed that the flux of θ is zero on the boundary $r = a$, which requires $\partial\theta/\partial r = 0$ there. Subtracting (4.34) from (4.35) we obtain the equation on $\theta' \equiv \theta - \bar{\theta}$:

$$\frac{\partial\theta'}{\partial t} + \frac{\partial(v\theta' - \overline{v\theta'})}{\partial x} + (v - \bar{v})\frac{\partial\bar{\theta}}{\partial x} = \kappa\Delta\theta'\,. \tag{4.36}$$

So far it is exact, now approximations start. At times far exceeding a^2/κ, the diffusion across the channel makes $\theta' \ll \bar{\theta}$, so that we can neglect in (4.36) all the terms containing θ' except $\kappa\Delta_r\theta'$ because nonuniform flow creates inhomogeneity across the pipe. Indeed, solving

$$(v - \bar{v})\frac{\partial\bar{\theta}}{\partial x} = \kappa\frac{1}{r}\frac{d}{dr}r\frac{d\theta'}{dr}\,. \tag{4.37}$$

with zero derivatives on the boundary and zero mean, we obtain

$$\theta'(x,r,t) = \left[6r^2 - 2 - 3(r/a)^4\right]\frac{\bar{v}}{24\kappa}\frac{\partial\bar{\theta}(x,t)}{\partial x}\,, \tag{4.38}$$

Using this solution one can explicitly check that the terms neglected in passing from (4.36) to (4.37) are small. The first term can be estimated as $\theta'/t \simeq \bar{v}a^2\bar{\theta}_x/\kappa t$, which is much smaller than $\bar{v}\bar{\theta}_x$, because $t \gg a^2\kappa$. The second term can be estimated as $\bar{v}\theta'_x \simeq (\bar{v}a)^2\bar{\theta}_{xx}/\kappa$, whose ratio with $\bar{v}\bar{\theta}_x$ is $\kappa\bar{\theta}_x/\bar{\theta}_{xx}\bar{v}a^2 \ll 1$. Indeed, $L \equiv \bar{\theta}_x/\bar{\theta}_{xx}$ is the longitudinal extent of the region filled by θ; stretching in nonuniform flow during the time t makes $L \simeq \bar{v}t \gg \bar{v}a^2/\kappa$.

We now substitute (4.38) into (4.35) and obtain

$$\frac{\partial \bar{\theta}}{\partial t} + \bar{v}\frac{\partial \bar{\theta}}{\partial x} = \left(\kappa + \frac{\bar{v}^2 a^2}{48\kappa}\right)\frac{\partial^2 \bar{\theta}}{\partial x^2} . \tag{4.39}$$

We thus see that the effective diffusivity is $\kappa(1 + Pe^2/192)$, where $Pe = 2\bar{v}a/\kappa$, as defined in Section 2.2.4.

2.4 Consider the continuity of the fluxes of mass, normal momentum and energy:

$$\rho_1 w_1 = \rho_2 w_2 , \quad P_1 + \rho_1 w_1^2 = P_2 + \rho_2 w_2^2 , \tag{4.40}$$

$$W_1 + w_1^2/2 = \frac{\rho_2 w_2}{\rho_1 w_1}\left(W_2 + w_2^2/2\right) = W_2 + w_2^2/2 . \tag{4.41}$$

Excluding w_1, w_2 from (4.40),

$$w_1^2 = \frac{\rho_2}{\rho_1}\frac{P_2 - P_1}{\rho_2 - \rho_1} , \quad w_2^2 = \frac{\rho_1}{\rho_2}\frac{P_2 - P_1}{\rho_2 - \rho_1} , \tag{4.42}$$

and substituting it into the Bernoulli relation (4.41) we derive the relation called the shock adiabat:

$$W_1 - W_2 = \frac{1}{2}\left(P_1 - P_2\right)(V_1 + V_2) . \tag{4.43}$$

For given pre-shock values of P_1, V_1, it determines the relation between P_2 and V_2. The shock adiabat is determined by two parameters, P_1, V_1, as distinct from the constant-entropy (Poisson) adiabate $PV^\gamma = \text{const.}$, which is determined by a single parameter, entropy. Of course, the after-shock parameters are completely determined if all the three pre-shock parameters, P_1, V_1, w_1, are given.

Substituting $W = \gamma P/\rho(\gamma - 1)$ into (4.43) we obtain the shock adiabat for a polytropic gas in two equivalent forms:

$$\frac{\rho_2}{\rho_1} = \frac{\beta P_1 + P_2}{P_1 + \beta P_2} , \quad \frac{P_2}{P_1} = \frac{\rho_1 - \beta \rho_2}{\rho_2 - \beta \rho_1} , \quad \beta = \frac{\gamma - 1}{\gamma + 1} . \tag{4.44}$$

Since pressures must be positive, the density ratio ρ_2/ρ_1 must not exceed $1/\beta$ (four and six for monatomic and diatomic gases, respectively). If the pre-shock velocity w_1 is given, the dimensionless ratios ρ_2/ρ_1, P_2/P_1 and $\mathcal{M}_2 = w_2/c_2 = w_2\sqrt{\rho_2/\gamma P_2}$ can be expressed via the dimensionless Mach number $\mathcal{M}_1 = w_1/c_1 = w_1\sqrt{\rho_1/\gamma P_1}$ by combining (4.42) and (4.44):

$$\frac{\rho_1}{\rho_2} = \beta + \frac{2}{(\gamma + 1)\mathcal{M}_1^2} , \quad \frac{P_2}{P_1} = \frac{2\gamma\mathcal{M}_1^2}{\gamma + 1} - \beta , \quad \mathcal{M}_2^2 = \frac{2 + (\gamma - 1)\mathcal{M}_1^2}{2\gamma\mathcal{M}_1^2 + 1 - \gamma} . \tag{4.45}$$

To have a subsonic flow after the shock, $\mathscr{M}_2 < 1$, one needs a supersonic flow before the shock, $\mathscr{M}_1 > 1$.

Thermodynamic inequality $\gamma > 1$ guarantees the regularity of all the above relations. The entropy is determined by the ratio P/ρ^γ; it is actually proportional to $\log(P/\rho^\gamma)$. Using (4.45) one can show that $s_2 - s_1 \propto \ln(P_2\rho_1^\gamma/P_1\rho_2^\gamma) > 0$, which corresponds to an irreversible conversion of the mechanical energy of the fluid motion into the thermal energy of the fluid.

See Chapters 85 and 89 of [16] for more details.

2.5 **Simple estimate.** We use a single shock, which has the form $u = -v\tanh(vx/2v)$ in the reference frame with zero mean velocity. We then simply get $\langle u^2 u_x^2 \rangle = 2v^5/15L$, so that

$$\varepsilon_4 = 6v[\langle u^2 u_x^2 \rangle + \langle u^2 \rangle \langle u_x^2 \rangle] = 24v^5/5L .$$

Substituting $v^5/L = 5\varepsilon_4/24$ into $S_5 = -32v^5 x/L$, we get

$$S_5 = -20\varepsilon_4 x/3 = -40vx[\langle u^2 u_x^2 \rangle + \langle u^2 \rangle \langle u_x^2 \rangle] . \qquad (4.46)$$

Sketch of a theory. One can also derive the evolution equation for the structure function, analogous to (2.10) and (2.43). Consider

$$\partial_t S_4 = -(3/5)\partial_x S_5 - 24v[\langle u^2 u_x^2 \rangle + \langle u_1^2 u_{2x}^2 \rangle] + 48v\langle u_1 u_2 u_{1x}^2 \rangle + 8v\langle u_1^3 u_{2xx} \rangle .$$

Since the distance x_{12} is in the inertial interval, we can neglect $\langle u_1^3 u_{2xx} \rangle$ and $\langle u_1 u_2 u_{1x}^2 \rangle$, and we can put $\langle u_1^2 u_{2x}^2 \rangle \approx \langle u^2 \rangle \langle u_x^2 \rangle$. Assuming that

$$\partial_t S_4 \simeq S_4 u/L \ll \varepsilon_4 \simeq u^5/L,$$

we neglect the lhs and obtain (4.46). Generally, one can derive

$$S_{2n+1} = -4\varepsilon_n x \frac{2n+1}{2n-1}.$$

2.6 Equations of motion are different for the cases of body mass either growing due to condensation or decreasing due to evaporation. In the latter case, we assume that the material is released with the body velocity u, so that evaporation by itself does not change $u(t)$. The velocity change then comes from the force acting on the body, so that the velocity time derivative is equal to the force divided by mass. On the contrary, the inertial force (1.34) is the time derivative of the quasimomentum. The equation of motion is then can be deduced from (1.36) and (4.49):

$$\rho_0 V(t)\dot{u} = \rho V(t)\dot{v} + \frac{\mathrm{d}}{\mathrm{d}t}\rho V(t)\frac{v-u}{2} . \qquad (4.47)$$

Since $\rho_0 \gg \rho$ then, comparing to v, we can neglect the body velocity u (but not its acceleration $\dot{u}$, of course):

$$2\rho_0 \dot{u} = 3\rho \dot{v} - \rho \alpha v |v| \ .$$

During the half period when $v = a \sin \omega t > 0$, the solution is

$$u(t) = \frac{3a\rho}{2\rho_0} \sin \omega t - \frac{a^2 \alpha \rho}{8\omega} (2\omega t - \cos 2\omega t) \ .$$

It shows that the volume decrease causes a negative drift.

2.7 **A rough estimate** can be obtained even without a proper understanding of the phenomenon. The effect must be independent of the phase of oscillations, i.e. of the sign of A; therefore, the dimensionless parameter A^2 must be expressed via the dimensionless parameter $P_0/\rho gh$. When the ratio $P_0/\rho gh$ is small we expect the answer to be independent of it, i.e. the threshold to be of order unity. When $P_0/\rho gh \gg 1$, the threshold must be large as well, since large P_0 decreases any effect of bubble oscillations, so one may expect the threshold at $A^2 \simeq P_0/\rho gh$. One can make a simple interpolation between the limits

$$A^2 \simeq 1 + \frac{P_0}{\rho gh} \ . \tag{4.48}$$

Qualitative explanation of the effect invokes compressibility of the bubble (Bleich 1956). Vertical oscillations of the vessel cause periodic variations of the gravitational acceleration. Upward acceleration of the vessel causes downward gravity, which provides for the buoyancy force directed up and vice versa for another half period. It is important that related variations of the buoyancy force do not average to zero since the volume of the bubble oscillates too because of oscillations of pressure due to the column of liquid above. The volume is smaller when the vessel accelerates upward since the effective gravity and pressure are then larger. As a result, the buoyancy force is lower when the vessel and the bubble accelerate upwards. The net result of symmetric up–down oscillations is thus a downward force acting on the bubble. When that force exceeds the upward buoyancy force provided by the static gravity g, the bubble sinks.

Theory. Consider an ideal fluid where there is no drag. The equation of motion in the vessel's reference frame is obtained from (1.35) by adding buoyancy and neglecting the mass of the air in the bubble:

$$\frac{\mathrm{d}}{2\mathrm{d}t} V(t) u = V(t) G(t), \quad G(t) = g + \ddot{x} \ . \tag{4.49}$$

Here $V(t)$ is the time-dependent bubble volume. Denote by z the bubble's vertical displacement with respect to the vessel, so that $u = \dot{z}$, positive upward. Assume the compressions and expansions of the bubble to be adiabatic, which requires the frequency to be larger than thermal diffusivity κ divided by the bubble size a. If, on the other hand, the vibration frequency is much smaller than the eigenfrequency (4.16) (sound velocity divided by the bubble radius) then one can relate the volume $V(t)$ to the pressure and the coordinates at the same instant of time:

$$PV^{\gamma}(t) = [P_0 + \rho G(h - z)]V^{\gamma} = (P_0 + \rho gh)V_0^{\gamma}.$$

Assuming small variations in z and $V = V_0 + \delta V \sin(\omega t)$ we get

$$\delta V = V_0 \frac{A\rho gh}{\gamma(P_0 + \rho gh)} \ . \tag{4.50}$$

The net change of the bubble momentum during the period can be obtained by integrating (4.49):

$$\int_0^{2\pi/\omega} V(t')G(t')\,dt' = \frac{2\pi V_0 g}{\omega}(1 - \delta VA/2V_0) + o(A^2). \tag{4.51}$$

The threshold corresponds to zero momentum transfer, which requires $\delta V = 2V_0/A$. According to (4.50), that gives the following answer:

$$A^2 = 2\gamma\left(1 + \frac{P_0}{\rho gh}\right) \ . \tag{4.52}$$

At this value of A, (4.49) has an oscillatory solution $z(t) \approx -(2Ag/\omega^2)$ $\sin(\omega t)$, which is valid when $Ag/\omega^2 \ll h$. Another way to interpret (4.52) is to say that it gives the depth h where small oscillations are possible for a given amplitude of vibrations A. A moment's reflection tells us that these oscillations are unstable, i.e. bubbles below h have stronger downward momentum transfer and will sink while bubbles above rise.

Notice that the threshold value does not depend on the frequency and the bubble radius (under an implicit assumption $a \ll h$). However, neglecting viscous friction is justified only when the Reynolds number of the flow around the bubble is large: $a\dot{z}/v \simeq aAg/\omega v \gg 1$, where v is the kinematic viscosity of the liquid. A different treatment is needed for small bubbles where inertia can be neglected compared with viscous friction and (4.49) is replaced by

$$4\pi va(t)\dot{z} = V(t)G(t) = 4\pi a^3(t)G/3 \ . \tag{4.53}$$

Here we used (4.21) for the viscous friction of a fluid sphere with the interchange water $\leftrightarrow$ air. Dividing by $a(t)$ and integrating over a period we get the velocity change proportional to $1 - \delta aA/a = 1 - \delta VA/3V_0$. Another difference is that $a^2 \ll \kappa/\omega$ for small bubbles, so that heat exchange is fast and we must use the isothermal rather than the adiabatic equation of state, i.e. put $\gamma = 1$ in (4.50). That gives the threshold which is again independent of the bubble size:

$$A^2 = 3 \left(1 + \frac{P_0}{\rho gh} \right) . \tag{4.54}$$

As we see, the rough estimate (4.48) is good for all cases.

2.8 **The answer is blowing in the wind.** The first explanation that comes to mind is that for upwind propagation the distance is larger by the factor $1 + v/c$ where v is the wind speed. Viscous dissipation in the air decreases the intensity q with the distance r according to $q \propto \exp(-v\omega^2 r/c^3)$. Another factor is spreading of the acoustic energy flux over a half-sphere of increasing radius: $q \propto r^{-2}$. Note, however, that near the ground (where most of our shouting happens), the wind speed is generally less than $15\,\mathrm{m\,s^{-1}}$ so that $v/c < 0.05$ and can be treated as a small factor. Therefore, if the difference in the distances traveled was the reason, the difference between upwind and downwind intensities would be of order v/c, i.e. small. Incidentally, there is also the Döppler shift with a relative change of frequency v/c, which is negligible (wind cannot turn tenor into bass).

The real reason for the fast intensity drop upwind is the inhomogeneous vertical profile of the horizontal wind speed near the ground. This bends sound rays, which encounter stronger winds when they are further from the ground. As shown in the figure, the refraction makes the spreading of acoustic rays near the ground much faster upwind than downwind. For sound propagating upwind at some angle to the ground, the part of the constant-phase surface that is higher moves slower, which turns the surface up (a similar phenomenon is also known from optics where rays bend toward a more optically dense medium). Such refraction creates a so-called "zone of silence" upwind from the source. Note how qualitatively different are upwind and downwind rays: eventually the former all go up while the latter all hit the ground.

To estimate the distance to the zone of silence, consider the rays propagating almost horizontally i.e. perpendicular to the velocity gradient. The wave front turns and rays bend with the angular velocity $\omega = dv/dz \simeq v/h$, where h is the height. The radius of curvature then is

$R = c/\omega = hc/v$. The zone of silence starts at the point where the ray hits the ground horizontally as shown in the figure. The distance to this point is $\ell = \sqrt{R^2 - (R-h)^2} \approx \sqrt{2Rh} = h\sqrt{2c/v}$. That formula disabuses one of the idea that it is better to be closer to the ground where the wind speed is less. On the contrary, taller person is heard further away. For $h = 1.5\,\text{m}$ and $v/c = 0.05$, the person of the same height will stop hearing you at the distance $2\ell \simeq 19\,\text{m}$ upwind. The boundary of the zone of silence is a caustic. See more in Section 4.3 of [21].

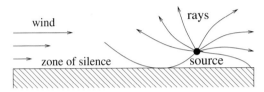

Similarly, the temperature gradient in the atmosphere causes the upward curvature of sound rays (now isotropic in the horizontal plane). The speed of sound is proportional to the square root of temperature in kelvin: $c(z) = c\sqrt{1 - \alpha z/T}$. Now the rays bend with $\omega = dc/dz = c\alpha/2T$, so that their radius of curvature is $R = 2T/\alpha$ and the radius of the silence zone is $\ell = 2\sqrt{Th/\alpha}$. As was mentioned in Section 1.1.3, $\alpha = 6.5$ kelvin per kilometer. Taking on the ground $T = 300$ and $c = 340\,\text{m}\,\text{s}^{-1}$, we obtain for $h = 10\,\text{m}$ the radius of the silence zone $\ell \simeq 1300\,\text{m}$ - that far one needs to go away to stop hearing annoying noise. To tell the truth, some noise penetrates the "zone of silence" because of diffraction, scattering on the ground and on wind turbulence.

On the contrary, when the ground cools during night producing local inversion of the vertical temperature profile, the rays curve downward and with multiple reflections from the ground can propagate quite far.

2.9 Let us call z the axis of symmetry and R the distance from this axis, so that the distance from the origin is $r = \sqrt{R^2 + z^2}$. One can consider a general spherical solution (2.27) as depending on both R, z and obtain the solution of the wave equation that depends only on R by integrating over z from 0 to ∞:

$$\phi(R,t) = \int_0^\infty dz\, r^{-1}[f_1(ct-r)+f_2(ct+r)] = \int_R^\infty \frac{f_1(ct-r)+f_2(ct+r)}{\sqrt{r^2-R^2}}\,dr$$

$$= \int_0^\infty [f_1(ct-R\cosh u)+f_2(ct+R\cosh u)]\,du\,. \qquad (4.55)$$

Here we used $dz = r\,dr/\sqrt{r^2-R^2}$ and $r = R\cosh u$. We saw in Section 2.3.1 that a plane sound wave in an ideal fluid propagates changing neither amplitude nor form, a spherical wave keeps the form but have the amplitude decreasing as an inverse radius. We see from (4.55) that a cylindrical wave keeps neither form nor shape. Another important distinction is that cylindrical wave may have a sharp front but not a sharp end: if the functions f_1, f_2 are localized, the large-time asymptotic decay at every point is $\phi \propto 1/ct$. This general difference in sound propagation in spaces of even and odd dimensionalities can be understood as follows. Let us pass in d-dimensional case from the wave equation to the Laplace equation in $d+1$ dimensions by the transform $t = \imath x_{d+1}/c$. The solution of the respective Poisson equation with the source is the integral $\int \dots d^{d+1}x' |\mathbf{x}-\mathbf{x}'|^{1-d}$, which has branch points for even d and poles for odd d. The large-time asymptotic of the integral is determined by the closest singularity: a pole gives exponential decay, while a branch-point gives power-law decay with time. If we lived in even-dimensional space, we would hear very different music, particularly drums. Taking $f_1(x) = \delta(x)$ and $f_2 = 0$, we obtain from (4.55) two-dimensional "boom": $\phi(R,t) = (c^2t^2 - R^2)^{-1/2}$ for $R < ct$ and $\phi(R,t) \equiv 0$ for $R > ct$. More on 2d is in Sections 70–72 of [16] and Art. 302 of [15].

2.10 Opposite signs of frequencies emitted and received means that the time runs in opposite directions for emitter and receiver: sound emitted later comes to receiver earlier. This way you can hear your music theme backwards (retrograde).

2.11 Since entropy is constant along a streamline in a steady flow of an ideal fluid, then one can express the pressure change via the density change as $dp = c^2 d\rho$, where $c^2 = (\partial p/\partial \rho)_s$. The steady Euler equation $\rho(\mathbf{v}\cdot\nabla)\mathbf{v} = -\nabla p = -c^2\nabla\rho$ along a streamline takes the form $\rho v\, dv = -c^2 d\rho$, that is density and velocity change in opposite ways along a steady flow. Indeed, negative/positive pressure gradient due to density gradient accelerates/decelerates the flow. We now use $d\rho/dv = -\rho v/c^2$, which can be also obtained from (2.31) in Section 2.3.2, to derive the flux-velocity relation:

$$\frac{d\rho v}{dv} = \rho\left(1 - \frac{v^2}{c^2}\right)\,. \qquad (4.56)$$

Therefore, the flux increases with the speed for subsonic regime and decreases for supersonic. Alternatively, one can ask what happens to velocity and density when the flux along a streamline increases (say, for converging flow). Denoting $M = v/c$ one can rewrite (4.56) as

$$d\ln v = \left(1 - M^2\right)^{-1} d\ln\rho v, \; d\ln\rho = -M^2 \left(1 - M^2\right)^{-1} d\ln\rho v. \quad (4.57)$$

For $M < 1$, the relative density change is M^2 times smaller than that of velocity, so that the flux increases at the expense of velocity, while density decreases. On the contrary, density change dominates supersonic regime where density increases and velocity decreases with the flux.

This property is exploited in the design of transonic nozzles: they are pinched in the middle (Laval 1888). In the converging part, the flux along every streamline increases to keep the total flux. The flow is subsonic in that part so that the flux increase is accompanied by the velocity increase. If the velocity does not reach the local speed of sound in the throat, where the cross-section area is minimal, then the flow decelerates in the expanding part. If, however, M does reach unity, it can happen only in the throat, where $d\ln\rho v = 0$, so that the right-hand sides of (4.57) are finite and the solution is smooth. The cross-section area starts expanding after the throat, so that the flux for every streamline decreases, which further accelerates the (now supersonic) flow. Of course, one cannot exceed the limiting velocity for the escape into vacuum, $c\sqrt{2/(\gamma-1)}$, derived in Section 1.1.4. Converging–diverging nozzle is used in rocket engines and many other devices.

2.12 We have seen in Exercise 1.2 that gravity cannot hold a finite mass of a gas at constant temperature in a vacuum. To describe a radial flow of such a gas, express pressure via density as $p = \rho RT$ and substitute the continuity equation $\rho v \propto r^{-2}$ into the steady Euler equation $\rho(\mathbf{v} \cdot \nabla)\mathbf{v} = -RT\nabla\rho - GM\rho/r^2$:

$$\left(v - \frac{RT}{v}\right)\frac{dv}{dr} = \frac{2RT}{r} - \frac{GM}{r^2}. \quad (4.58)$$

We see that RT plays the role of isothermal squared speed of sound. Without gravity, the right hand side is positive and the flow behaves according to the general rule described in Exercise 2.11: since the flux ρv decreases along every streamline, then subsonic flow slows down while supersonic flow accelerates with the distance. The gravity allows for an accelerating transonic flow, similar to that in the nozzle from Exercise 2.11, if the star radius is less than $r_* = GM/2RT$. Then, the smooth solution of (4.58) is a subsonic flow at $r < r_*$, which accelerates to $v(r_*) = \sqrt{RT}$, and then continues to accelerate in a supersonic regime

at $r > r_*$, asymptotically $v(r) = 2\sqrt{RT \log(r/r_*)}$ (Parker 1958).
Acceleration in the subsonic regime is due to the decrease of
gravitational energy, while acceleration at large distances is due to
pressure gradient caused by the decrease of density in the supersonic
regime. Note that for isothermal flow there is no limit for the velocity of
the escape into vacuum since $\gamma = 1$, yet physically at some distance
thermal conduction fails to keep the temperature constant.

4.3 Chapter 3

3.1 The wave velocity, $v_g = \sqrt{gh}$, increases with the depth. Depth usually
decreases as one approaches the shore. Therefore, even if the wave
comes from the deep at an angle, the parts of the wavefronts that are in
deeper water move faster changing the orientation of the fronts (similar
to refraction in Exercise 2.8).

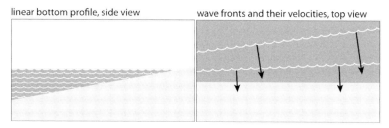

linear bottom profile, side view   wave fronts and their velocities, top view

3.2 The length of the wave packet is $L = N\lambda = 2\pi N/k$. The wave packet
propagates with v_{group}. The float will be disturbed by the wave packet
during the time: $\tau = L/v_{\text{group}} = 2\pi N/k v_{\text{group}}$. For a quasimonochromatic
wave packet $T = 2\pi/\omega(k)$, the number of "up and down" motions of the
wave packet is $n = \tau/T = N v_{\text{phase}}/v_{\text{group}}$. For gravity waves on deep
water $\omega = \sqrt{gk}$ we have $n = 2N$. For gravity-capillary waves on deep
water $\omega = \sqrt{(gk + \sigma k^3/\rho)}$ we have

$$n = 2N \frac{(gk + \sigma k^3/\rho)}{k(gk + 3\sigma k^2/\rho)}$$

and for purely capillary waves $\omega = \sqrt{\sigma k^3/\rho}$, $n = 2N/3$.

Remark. This was a linear consideration (due to [31]). Nonlinear
Stokes drift causes the float to move with the wave so that the relative
speed is less: $v_{\text{group}} \to v_{\text{group}}[1 - (ak)^2]$.

3.3 **Qualitative analysis and simple estimate.** The group velocity of surface waves depends on their wavelength nonmonotonically as shown in the figure:

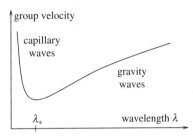

The stone excites waves with wavelengths not less than its size. Indeed, a Fourier image of a bump of width l is nonzero for wavenumbers less than $1/l$. Therefore, the wave pattern on a water surface depends on the relation between the size of the initial perturbation and $\lambda_* = 2\pi/k_*$. If we drop a large stone (with a size far exceeding λ_*) it will mostly generate gravity waves which are faster when longer. As a result, the wavelength increases with the distance from the origin, that is, the larger circles are progressively sparser. On the contrary, small stones and raindrops can also generate capillary waves with the shortest being fastest so that the distance between circles decreases with the radius. Since there is a minimal group wave speed $v_* \simeq (g\alpha/\rho)^{1/4} \simeq 18\,\mathrm{cm\,s^{-1}}$, there are no waves inside the circle with the radius $v_* t$; this circle is a caustic. According to Section 3.1.4, the perturbation amplitude on the caustic decays as $t^{-5/6}$ in a two-dimensional case. Remind that for water waves the minimal group velocity v_* is less than the minimal phase velocity, in particular, $v_* < \omega_*/k_*$ so that all wave crests emerge out of nowhere on the caustic and run forward from it. A detailed consideration of caustics, including stone-generated surface waves can be found in [17], Section 4.11.

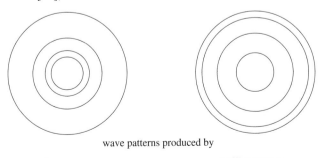

wave patterns produced by

gravity waves   capillary waves

Figure 4.2 Long gravity waves and short capillary ripples propagating together. Photograph copyright: Cammeraydave, www.dreamstime.com.

Both gravity and capillary waves can be seen in Figure 4.2. For a given speed exceeding v_*, there are two distinct wavelengths that propagate together.

A sketch of a theory for gravity waves is based on Sections 3.1.1 and 3.1.4. The perturbation can be considered as a force localized in both space and time. Since we aim to describe time intervals far exceeding the time of the stone sinking and wavelengths far exceeding its size we model this force by the product of delta functions $\delta(t)\delta(\mathbf{r})$. Adding such a force to the equation of motion (3.4) we get

$$\frac{\partial v_z}{\partial t} = \frac{\partial^2 \zeta}{\partial t^2} = -g\frac{\partial \zeta}{\partial z} + \delta(t)\delta(\mathbf{r})$$

which gives $\zeta_k = \mathrm{i}\exp(-\mathrm{i}\sqrt{gk}\,t)/\sqrt{gk}$ and

$$\zeta(r,t) = \frac{\mathrm{i}}{8\pi^2\sqrt{g}} \int \sqrt{k}\,dk\,d\theta \exp(\mathrm{i}kr\cos\theta - \mathrm{i}\sqrt{gk}\,t) \tag{4.59}$$

$$= \frac{\mathrm{i}}{4\pi^2\sqrt{g}} \int_0^\infty dk\sqrt{k}J_0(kr)\exp(-\mathrm{i}\sqrt{gk}\,t) \equiv \frac{\mathrm{i}}{2\pi g^2 t^3}\Phi\left(\frac{r}{gt^2}\right),$$

$$\Phi(y) = \int_0^\infty dz z^2 J_0(yz^2)\exp(-\mathrm{i}z) . \tag{4.60}$$

Here J_0 is the Bessel function. One can see that the crests accelerate with the acceleration due to gravity g. Before the leading crest, for $r \gg gt^2$, the surface is steady:

$$\zeta(r) \propto g^2 t^{-3} (gt^2/r)^{3/2} \propto g^{-1/2} r^{-3/2} .$$

Behind the leading crest, for $r \ll gt^2$, the main contribution to (4.59) is given by $\theta = 0$, $r = \omega'(k)t = t\sqrt{g/k}/2$, that is by $k = gt^2/4r^2$, so that

$$\zeta(r,t) \propto \sin(gt^2/4r) .$$

The crests are at the following radial positions:

$$r_n = \frac{gt^2}{2\pi(4n+1)} .$$

Maximal speed of the gravity waves is determined by the fluid depth h, so that there are no waves outside the circle with the radius $t(gh)^{1/2}$, which is another caustic. For details, see www.tcm.phy.cam.ac.uk/~dek12/, Tale 8 by D. Khmelnitskii. A general consideration that accounts for a finite stone size and capillary waves can be found in section 17.09 of [14].

3.4 The Hamiltonian is real, therefore $A(k)$ is real; we denote A_1, A_2, respectively, as its symmetric and antisymmetric parts. Also, $B(k) = B(-k)$ and we can consider $B(k)$ real, absorbing its phase into $b(k)$. The same is true for the transformation coefficients u, v, which also could be chosen real (as is also clear below). The canonicity of u–v transformation requires

$$u^2(k) - v^2(k) = 1, \quad u(k)v(-k) = u(-k)v(k)$$

and suggests the substitution

$$u(k) = \cosh[\zeta(k)], \quad v(k) = \sinh[\zeta(k)] .$$

In these terms, the transformation equations take the form

$$\omega_k = A_2 + A_1/\cosh(2\zeta), \quad A_1 \sinh(2\zeta) = B[\cosh^2(\zeta) + \sinh^2(\zeta)] .$$

That gives

$$\omega_k = A_2(k) + \operatorname{sign} A_1(k)\sqrt{A_1^2(k) - B^2(k)} .$$

If $A_1(k)$ turns into zero when $B(k) \neq 0$, we have a complex frequency, which describes exponential growth of waves, i.e. instability. In this

case, the quadratic Hamiltonian cannot be written in the standard form
(3.32), but it can be reduced to the form

$$\mathscr{H}_2 = \int C(k)[a(k)a(-k) + a^*(k)a^*(-k)]\,d\mathbf{k},$$

which describes the creation of the pair of quasiparticles from vacuum
and the inverse process, see [34], section 1.1.

3.5 For two-dimensional wavevectors, $\mathbf{k} = \{k_x, k_y\}$, the function $\omega(\mathbf{k})$
determines a surface in the three-dimensional space ω, k_x, k_y. In an
isotropic case, $\omega(k)$ determines a surface of revolution. Consider two
such surfaces: S determined by $\omega(k)$ and S_1 by $\omega(k_1)$. The possibility of
finding $\mathbf{k}_1, \mathbf{k}$ such that $\omega(\mathbf{k}_1 + \mathbf{k}) = \omega(k_1) + \omega(k)$ means that the second
surface must be shifted up by $\omega(k)$ and right by $\mathbf{k}$ and that the two
surfaces must intersect:

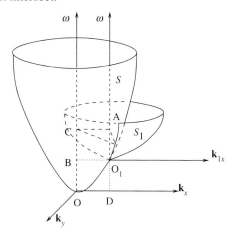

All three points with the coordinates $\{\omega, k\}$, $\{\omega_1, k_1\}$ and $\{\omega(k + k_1),$
$\mathbf{k} + \mathbf{k}_1\}$ must lie on the intersection line O_1A. For example, $OD = k$,
$OB = \omega$, $BC = \omega_1$. For intersection, $\omega(k)$ must define a convex surface.
For a power law, $\omega(k) \propto k^\alpha$, this requires $\alpha > 1$. For $\alpha = 1$, only waves
with collinear wavevectors can interact resonantly.

3.6 The equation for the standing soliton analogous to (3.45) is now

$$\omega'' \Delta A = 2T(A^3 - A_0^2 A).$$

In polar coordinates, $\Delta = r^{-1}\partial_r r \partial_r + r^{-2}\partial_\phi^2$. It is thus clear that the
dependence on the angle ϕ makes the condensate amplitude become
zero on the axis: $A \propto r$ as $r \to 0$. Denote $f(r/r_0) \equiv A/A_0$. Considering
$r \to \infty$ one concludes that $r_0^2 = \omega''/(2TA_0^2)$, i.e. the amplitude at infinity

and the size are inversely related, as in 1D. The resulting equation in the variable $\xi = r/r_0$ has the form

$$\frac{1}{\xi}\frac{d}{d\xi}\left(\xi\frac{df}{d\xi}\right) + \left(1 - \frac{1}{\xi^2}\right)f - f^3 = 0, \qquad (4.61)$$

This can be solved numerically. When $\xi \to 0$ then $f \propto \xi$. When $\xi \to \infty$, $f = 1 + \delta f$ and in a linear approximation we get

$$\frac{1}{\xi}\frac{d}{d\xi}\left(\xi\frac{d\delta f}{d\xi}\right) - \frac{1}{\xi^2} - 2\delta f - \frac{1}{\xi^2}\delta f = 0, \qquad (4.62)$$

which has a solution $\delta f = -1/2\xi^2 + O(\xi^{-4})$. The solution is a vortex because there is a line in space, $r = 0$, where $|\psi|^2 = 0$. Going around the vortex line, the phase acquires 2π. That this is a vortex is also clear from the fact that there is a current J (and the velocity) around it:

$$J_\phi \propto i\left(\psi\frac{\partial\psi^*}{r\partial\phi} - \psi^*\frac{\partial\psi}{r\partial\phi}\right) = \frac{A^2}{r},$$

so that the circulation is independent of the distance from the vortex. Note that the kinetic energy of a single vortex diverges logarithmically with the system size (or with the distance to another parallel vortex line with an opposite circulation): $\int v^2\,rdr \propto \int dr/r$. This has many consequences in different fields of physics, in particular in two dimensions (like the Berezinski–Kosterlitz–Thouless phase transition of vortex pairing).

3.7 **Qualitative answer**. To integrate a general system of $2n$ first-order differential equations one needs to know $2n$ conserved quantities. For a Hamiltonian system though, Liouville's theorem tells us that n independent integrals of motion are enough to make the system integrable and its motion equivalent to n oscillators. For three modes, the number of degrees of freedom is exactly equal to the number of the general integrals of motion: the Hamiltonian $\mathcal{H}$, the total number of waves $P = \sum|a_m|^2$ and the momentum $M = \sum m|a_m|^2$. This means that the system is integrable: in the six-dimensional space of a_0, a_1, a_{-1}, every trajectory lies on a three-torus.

Quantitative solution. The full system,

$$i\frac{da_0}{dt} = T\left(|a_0|^2 + 2|a_1|^2 + 2|a_{-1}|^2\right)a_0 + 2Ta_1a_{-1}a_0^*, \qquad (4.63)$$

$$i\frac{da_1}{dt} = \beta a_1 + T\left(|a_1|^2 + 2|a_0|^2 + 2|a_{-1}|^2\right)a_1 + Ta_{-1}^*a_0^2, \qquad (4.64)$$

$$i\frac{da_{-1}}{dt} = \beta a_{-1} + T\left(|a_{-1}|^2 + 2|a_0|^2 + 2|a_1|^2\right)a_0 + Ta_1^*a_0^2, \qquad (4.65)$$

can be reduced explicitly to a single degree of freedom using the integrals of motion. Consider for simplicity $M = 0$ and put $a_0 = A_0 \exp(i\theta_0)$, $a_1 = A_1 \exp(i\theta_1)$ and $a_{-1} = A_1 \exp(i\theta_{-1})$, where $A_0, \theta_0, A_1, \theta_1, \theta_{-1}$ are real. Introducing $\theta = 2\theta_0 - \theta_1 - \theta_{-1}$, the system can be written as:

$$\frac{dA_0}{dt} = -2\frac{dA_1}{dt} = -2TA_0A_1^2 \sin\theta \ , \tag{4.66}$$

$$\frac{d\theta}{dt} = 2\beta + 2T(A_0^2 - A_1^2) + 2T(A_0^2 - 2A_1^2)\cos\theta \ . \tag{4.67}$$

This system conserves the Hamiltonian $\mathcal{H} = 2\beta A_1^2 + TA_0^4/2 + 3TA_1^4 + 4TA_0^2A_1^2 + 2TA_1^2A_0^2\cos\theta$ and $P = A_0^2 + 2A_1^2$. Introducing $B = A_1^2$ one can write $\mathcal{H} = (2\beta + TP/2)B + TB(P - 2B)(2\cos\theta + 3/2)$ and

$$\frac{dB}{dt} = 2TB(P - 2B)\sin\theta = -\frac{\partial \mathcal{H}}{\partial \theta} \ , \tag{4.68}$$

$$\frac{d\theta}{dt} = 2\beta + 2T(P - 3B) + 2T(P - 4B)\cos\theta = \frac{\partial \mathcal{H}}{\partial B} \ . \tag{4.69}$$

Expressing θ via $\mathcal{H}, B$ and substituting into (4.68), one can find the solution in terms of an elliptic integral. To understand the qualitative nature of the motion, it helps to draw the contours of constant $\mathcal{H}$ in coordinates $B, \cos\theta$. The evolution proceeds along those contours. The phase space is restricted by two straight lines $B = 0$ (which corresponds to $\mathcal{H} = 0$) and $B = P/2$ (which corresponds to $A_0 = 0$ and $\mathcal{H} = \beta P + TP^2/4$).

It is instructive, in particular, to compare the phase portraits with and without the modulational instability. Indeed, exactly as in Section 3.3.2, we can consider $|a_0| \gg |a_1|, |a_{-1}|$ and $|a_0|^2 \approx P$, linearize (4.64, 4.65) and for $a_1, a_{-1} \propto \exp(i\Omega t - iTPt)$ obtain $\Omega^2 = \beta(\beta + 2TP)$ in agreement with (3.5). When $\beta T < 0$, there is an instability, while in the phase portrait of (4.68, 4.69) there exist two fixed points at $B = 0$, $\cos\theta_0 = -1 - \beta/TP$, which are saddles connected by a separatrix. Even for trajectories that start infinitesimally close to $B = 0$, the separatrix makes it necessary to deviate to finite B, as seen from the right panel of Figure 4.3.

In the Hamiltonian integrable case, the separatrix is an unstable manifold for one fixed point and a stable manifold for another. When the system is perturbed (say, by small pumping and damping) and loses its Hamiltonian structure and integrability, the stable manifold of one point does not coincide with the unstable manifold of another, but those two

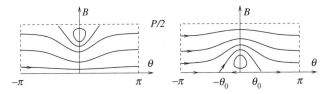

Figure 4.3 Sketch of the phase-space trajectories for large positive β (left, stable case) and $-2TP < \beta < 0$ (right, unstable case).

manifolds have an infinite number of intersections instead. This separatrix splitting is responsible for creating a chaotic attractor.

3.8 As in any shock, mass and momentum are conserved but the mechanical energy is not.

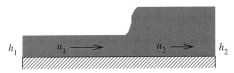

Mass flux is $\rho u h$ so that the flux constancy gives $u_1 h_1 = u_2 h_2$. The momentum flux (1.15) includes the pressure, which must be integrated over the vertical coordinate. Where the flow is uniform, the pressure can be taken to be hydrostatic, $P(z) = \rho g(h - z)$, which gives the momentum flux as follows:

$$\int_0^h \left[p(z) + \rho u^2 \right] dz = \rho g h^2 / 2 + \rho h u^2 .$$

The momentum flux constancy gives $g h_1^2 + 2 h_1 u_1^2 = g h_2^2 + 2 h_2 u_2^2$. Substituting here $u_2 = u_1 h_1 / h_2$ we find $g h_1^2 - g h_2^2 = g(h_1 - h_2)$ $(h_1 + h_2) = 2 u_1^2 h_1^2 / h_2 - 2 h_1 u_1^2 = 2 u_1^2 h_1 (h_1 - h_2)/h_2$. This gives the height after the jump:

$$\frac{h_2}{h_1} = -\frac{1}{2} + \sqrt{\frac{1}{4} + 2 \frac{u_1^2}{g h_1}} \approx 1 + \frac{2\varepsilon}{3} .$$

This is actually the first Rankine–Hugoniot relation (4.45) taken for $\gamma = 2$, $\mathcal{M}_1^2 = 1 + \varepsilon$.

The flux of mechanical energy is as follows

$$\int_0^h \left[p(z) + \rho u^2 / 2 + \rho g z \right] u \, dz = \rho g u h^2 + \rho h u^3 / 2 .$$

Here the first term is the work done by the pressure forces, the second term is the flux of the kinetic energy and the third term is the flux of the potential energy. The difference in energy fluxes before and after the front is then the energy dissipation rate:

$$\rho g u_1 h_1^2 + \rho h_1 u_1^3/2 - \rho g u_2 h_2^2 - \rho h_2 u_2^3/2$$

$$= \rho g u_1 h_1 (h_1 - h_2)$$

$$+ \frac{\rho u_1}{4}[2 h_2 u_2^2 + g(h_2^2 - h_1^2)] - \frac{\rho u_2}{4}[2 h_1 u_1^2 + g(h_1^2 - h_2^2)]$$

$$= \rho g u_1 h_1 (h_1 - h_2) + \frac{\rho(u_1 + u_2)}{4} g(h_2^2 - h_1^2)$$

$$= \rho g u_1 (h_2 - h_1)^3/4 h_2 \approx 2\varepsilon^3 \rho u_1^5/27 g \ . \tag{4.70}$$

Comparing (4.70) with the Rankine–Hugoniot energy–continuity relation (4.41) from Exercise 2.3, the difference is that for a gas we wrote the total energy, which is, of course, conserved. Another difference is that the pressure is determined by height (analogue of density) for a shallow fluid, so that two conservation laws of mass and momentum are sufficient in this case to determine the velocity and height after the shock. The dissipation rate is then completely determined by the preshock laminar flow via the conservation laws of the mass and momentum. To appreciate this amazing result, the reader is advised to make observations in the kitchen sink to see how complicated and turbulent the real hydraulic jump is. In particular, it dissipates energy by leaving behind dispersive waves with wavelengths comparable to the fluid depth which propagate slower (see [17], section 2.12 for more details).

3.9 For a running wave of the form $u(x - vt)$ the equation

$$u_t + u u_x + \beta u_{xxx} - \mu u_{xx} = 0$$

takes the form

$$-vu + u^2/2 + \beta u_{xx} - \mu u_x = \text{const.}$$

Integrate it using the boundary conditions and introduce $\tau = x\sqrt{v/\beta}$ and $q(\tau) = u/v$:

$$\ddot{q} = -2\lambda \dot{q} + q - q^2/2 \quad 2\lambda \equiv \mu/\sqrt{\beta v} \ . \tag{4.71}$$

This is a Newtonian equation for a particle in the potential $U = q^3/6 - q^2/2$ under the action of the friction force (we assume $\beta > 0$). The initial conditions in the distant past are $q(-\infty) = \dot{q}(-\infty) = \ddot{q}(-\infty) = 0$ so that the particle has zero energy: $E = \dot{q}/2 + q^3/6 - q^2/2$. We consider $q \geq 0$ since negative q goes to $q(\infty) = -\infty$, which gives nonphysical u. Friction eventually brings the particle to the minimum of the potential: $q \to 2$ as $\tau \to \infty$. Near the minimum we can use the harmonic approximation,

$$\ddot{q} \simeq -2\lambda\dot{q} - (q-2), \qquad (4.72)$$

which gives the law of decay: $q - 2 \propto \exp(rt)$ with $r_{1,2} = -\lambda \pm \sqrt{\lambda^2 - 1}$. We see that for $\lambda < 1$ the asymptotic decay is accompanied by oscillations while for $\lambda \geq 1$ it is monotonic. For $\lambda \ll 1$, an initial evolution is described by the (soliton) solution without friction,

$$q = \frac{3}{ch^2[(t - t_0)/2]},$$

which brings the particle almost to $q \approx 3$, after which it goes back almost to $q \approx 0$, and then approaches minimum, oscillating. On the other hand, for very large friction, $\lambda \gg 1$, the particle goes to the minimum monotonically and the solution is close to the Burgers shock wave. It is clear that there exists an interval of $\lambda > 1$ where the solution (called a collision-less shock) has a finite number of oscillations. It is an interesting question how large λ must be so that no oscillations are present.

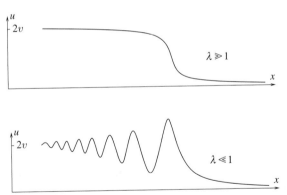

Note that adding a small dispersion does not change the shock much, while adding even small dissipation turns the soliton into the (collision-less) shock, i.e. strongly changes the whole $x \to -\infty$ asymptotics; this is because friction is a symmetry-breaking perturbation, as it breaks time-reversal symmetry.

3.10 **Simple estimate**. Let us first use the virial theorem to estimate the dispersion relation for $v = 0$. This is done following (3.1, 3.2). Both fluids are involved in the motion, so the kinetic energy per unit area can be estimated as $(\rho_1 + \rho_2)\omega^2 a^2 \lambda$, where a is the surface elevation. Under such elevation, the gravitational potential energy increases for the lower fluid and decreases for the upper fluid so that the net potential energy per unit area is $(\rho_2 - \rho_1)ga^2$. The potential energy of surface tension is the same $\alpha(a/\lambda)^2$. The virial theorem then generalizes (3.2) into

$$(\rho_1 + \rho_2)\omega^2 \simeq (\rho_2 - \rho_1)g\lambda^{-1} + \alpha\lambda^{-3}. \qquad (4.73)$$

Complete solution. We need to combine the approach of Section 2.1 with that of Section 3.1. We introduce the velocity potentials ϕ_1 and ϕ_2 on both sides of the surface. Then the respective values of the pressure in the approximation linear with respect to the potentials ϕ_1, ϕ_2 and the elevation ζ are:

$$P_1 = \rho_1 \left(g\zeta + \frac{\partial \phi_1}{\partial t} + v\frac{\partial \phi_1}{\partial x} \right), \quad P_2 = \rho_2 \left(g\zeta + \frac{\partial \phi_2}{\partial t} \right). \qquad (4.74)$$

The pressure difference between the two sides is balanced by the surface tension as in (3.16):

$$\rho_2 \left(g\zeta + \frac{\partial \phi_2}{\partial t} \right) - \rho_1 \left(g\zeta + \frac{\partial \phi_1}{\partial t} + v\frac{\partial \phi_1}{\partial x} \right) = \alpha \frac{\partial^2 \zeta}{\partial x^2}. \qquad (4.75)$$

We express the potential ϕ_1, ϕ_2 via the elevation ζ using the kinematic boundary conditions:

$$\frac{\partial \phi_2}{\partial z} = \frac{\partial \zeta}{\partial t}, \quad \frac{\partial \phi_1}{\partial z} = \frac{\partial \zeta}{\partial t} + v\frac{\partial \zeta}{\partial x}. \qquad (4.76)$$

For $\zeta(x,t) \propto \exp(ikx - i\Omega t)$ we obtain from (4.75 and 4.76) the dispersion relation

$$\Omega^2 - 2\frac{\rho_1}{\rho_1 + \rho_2}vk\Omega + \frac{\rho_1}{\rho_1 + \rho_2}v^2k^2 - c_0^2 k^2 = 0, \qquad (4.77)$$

$$c_0^2 = \frac{\rho_2 - \rho_1}{\rho_1 + \rho_2}\frac{g}{k} + \frac{\alpha k}{\rho_1 + \rho_2}. \qquad (4.78)$$

Here $k > 0$ and c_0 is the phase velocity of the gravity-capillary waves in the case $v = 0$. In this case, we see that when $\rho_1 > \rho_2$ the phase velocity and the frequency are imaginary for sufficiently long waves with $k^2 < (\rho_1 - \rho_2)g/\alpha$. This signals the so-called Rayleigh–Taylor instability, which is responsible, in particular, for water falling out of an upturned glass. In this case, inverted gravity causes the instability while surface tension stabilizes it.

Consider now a bottom-heavy configuration, $\rho_1 < \rho_2$, which is stable for $v = 0$ since $c_0^2 > 0$. For sufficiently high v, we have the Kelvin–Helmholtz instability, when one can find k such that the determinant of the quadratic equation (4.77) is negative:

$$v^2 > \frac{(\rho_1 + \rho_2)^2}{\rho_1 \rho_2} \min_k c_0^2(k) = \frac{\rho_1 + \rho_2}{\rho_1 \rho_2} \sqrt{4(\rho_2 - \rho_1)g\alpha} \ . \tag{4.79}$$

Again, as in the Landau criterion for generating excitations in a superfluid, the criterion for the growth of surface waves due to instability is the possibility that the flow moves faster than the minimal phase velocity of the waves. In this case, both gravity and surface tension are needed to provide for a nonzero minimal velocity; in other words, their interplay makes the flow stable for lower speeds. Application of (4.79) to wind upon water, where $\rho_1/\rho_2 \simeq 10^{-3}$ gives the threshold value

$$v_{\text{th}} \approx 30 \left(\frac{4g\alpha}{\rho_2} \right)^{1/4} \approx 7\,\text{m}\,\text{s}^{-1}.$$

This is unrealistically high. By blowing air over a cup of tea, it is easy to observe that one can ruffle the surface with much smaller air velocity. A realistic theory of wave generation by wind not only requires an account of viscosity but also of the fact that the wind is practically always turbulent; strong coupling between water waves and air vortices makes such a theory non-trivial.

3.11 As in Sections 2.3.1 and 3.1.1 we consider the (Lagrangian) equation for the coordinate of the fluid particle,

$$\frac{d\mathbf{R}}{dt} = \mathbf{v}(\mathbf{R}, t) = \mathbf{V}(\mathbf{R}) \sin(\omega t) + \mathbf{U}(\mathbf{R}) \sin(\omega t + \varphi) \ , \tag{4.80}$$

and solve it perturbatively, expanding in the inverse powers of ω. Assume without loss of generality $\mathbf{R}(0) = 0$, then in the first order

$$\mathbf{R}_1(t) = -\omega^{-1}[\mathbf{V}(0)\cos(\omega t) + \mathbf{U}(0)\cos(\omega t + \varphi)] \ . \tag{4.81}$$

We now substitute it into the equation for the second-order displacement:

$$\frac{d\mathbf{R}_2}{dt} = (\mathbf{R}_1 \cdot \nabla)[\mathbf{V}\sin(\omega t) + \mathbf{U}\sin(\omega t + \varphi)] \ . \tag{4.82}$$

Integration over period gives the net Lagrangian drift velocity

$$\int_0^{2\pi/\omega} \frac{d\mathbf{R}}{dt}\frac{dt\,\omega}{2\pi} = \frac{\sin\varphi}{2\omega}[(\mathbf{V}\cdot\nabla)\mathbf{U} - (\mathbf{U}\cdot\nabla)\mathbf{V}] \,. \qquad (4.83)$$

We see that a round trip in the space of Eulerian flows can produce a net Lagrangian displacement if two things are present: phase shift in time and misalignment in space. The former is represented in (4.83) by the factor $\sin\varphi$ and the latter by the square bracket, which contains what can be called the commutator of two vector fields. Several threads running through this book come together in this simple formula. First, it is a reminder of a nonlinear relation between Eulerian and Lagrangian velocities. Second, nonzero drift requires irreversibility whose degree can be measured by the area of a cycle in respective space, as we did for low-Reynolds swimmer in Section 1.5.1. Indeed, if horizontal and vertical displacements are respectively $\sin(\omega t)$ and $\sin(\omega t + \varphi)$, then the area of the cycle is $\sin\varphi$, so that φ is sometimes called geometrical phase. Note that even if the fields $\mathbf{V}(\mathbf{R}), \mathbf{U}(\mathbf{R})$ are potential, the net flow (4.83) generally has nonzero vorticity. In the particular case of incompressible flows in two dimensions, when $\mathbf{V} = z \times \nabla\psi_1$ and $\mathbf{U} = z \times \nabla\psi_2$, the net flow has a stream function ψ proportional to the vector product of two fields (again, the area of a triangle), that is to the Poisson bracket of the two stream functions:

$\psi = [\mathbf{V} \times \mathbf{U}]\sin\varphi/2\omega = (\psi_{1x}\psi_{2y} - \psi_{1y}\psi_{2x})\sin\varphi/2\omega.$

Epilogue

Now is a good time to reread the Prologue and make sure that we have learned basic mechanisms and elementary interplay between nonlinearity, dissipation and dispersion in fluid mechanics. Where can we go from here? It is important to recognize that this book describes only a few basic types of flow and leaves whole sets of physical phenomena outside of its scope. It is impossible to fit all of fluid mechanics into the format of a single story with a few memorable protagonists. Here is a brief guide to further reading; more details can be found in the endnotes.

A comparable elementary textbook (which is about twice as big) is that of Acheson [2]; it provides extra material and some alternative explanations on the subjects described in Chapters 1 and 2. On the subjects of Chapter 3, a timeless classic is the book by Lighthill [17]. For a deep and comprehensive study of fluid mechanics as a branch of theoretical physics one cannot do better than use another timeless classic, volume VI of the Landau–Lifshitz course [16]. Apart from a more detailed treatment of the subjects covered here, it contains a variety of different flows, a detailed presentation of the boundary layer theory, the theory of diffusion and thermal conductivity in fluids, relativistic and superfluid hydrodynamics, etc. In addition to reading about fluids, it is worth looking at flows, which is as appealing aesthetically as it is instructive and helpful in developing a physicist's intuition. Plenty of visual material, both images and videos, can be found in [13, 30] and the Galleries at www.efluids.com. And last but not least: the beauty of fluid mechanics can be revealed by simply looking at the world around us and trying simple experiments in a kitchen sink, bathtub, swimming pool, etc. It is likely that fluid mechanics is the last frontier where fundamental discoveries in physics can still be made in such a way.

After learning what fluid mechanics can do for you, some of you may be interested to know what you can do for fluid mechanics. Let me briefly mention several directions of present-day action in the field. Considerable analytical and numerical work continues to be devoted to the fundamental properties of the equations of fluid mechanics, particularly to the existence and uniqueness of solutions. The subject of a finite-time singularity in incompressible flows particularly stands out. These are not arcane subtleties of mathematical description but questions whose answers determine important physical properties, for instance, statistics of large fluctuations in turbulent flows. On the one hand, turbulence is a paradigmatic far-from-equilibrium state where we hope to learn general laws governing nonequilibrium systems; on the other hand, its ubiquity in nature and industry requires knowledge of many specific features. Therefore, experimental and numerical studies of turbulence continue toward both deeper understanding and wider applications in geophysics, astrophysics and industry; see e.g. [7, 12, 34]. At the other extreme, we have seen that flows at very low Reynolds numbers are far from trivial; the needs of biology and industry have triggered an explosive development of microfluidics, bringing new fundamental phenomena and amazing devices. Despite a natural tendency of theoreticians toward limits (of low and high $Re, Fr, \mathcal{M}$), experimentalists, observers and engineers continue to discover fascinating phenomena for the whole range of flow control parameters.

The domain of quantum fluids continues to expand, now including superfluid liquids, cold gases, superconductors, electron droplets and other systems. Quantization of vorticity and a novel factor of disorder add to the interplay of nonlinearity, dispersion and dissipation. Many phenomena in plasma physics also belong to a domain of fluid mechanics. Quantum fluids and plasma can often be described by two interconnected fluids (normal and superfluid, electron and ion), allowing for a rich set of phenomena.

Another thriving subject is the study of complex fluids. One important example is a liquid containing long polymer molecules that are able to store elastic stresses providing fluid with a memory. That elastic memory provides for inertia (and nonlinearity) of its own and introduces the new dimensionless control parameter, the Weissenberg number, which is the product of fluid strain and the polymer response time. When the Weissenberg number increases, symmetries are broken and instabilities take place even at very low Reynolds numbers. For example, left–right asymmetry of the flow past a body appears because polymers remain stretched for some time after leaving the high-strain region near the body (compare with Section 1.5.3). Similarly, instability may appear when flow perturbation changes elastic stresses in a way that reinforces perturbation, culminating in so-called elastic turbulence [27]. Another example is a two-phase flow with numerous applications, from clouds to internal combustion engines; here, a lot of interesting physics is related to the relative inertia of two phases and a very inhomogeneous distribution of droplets, particles or bubbles in a flow.

And coming back to basics: our present understanding of how fish and micro-organisms swim and how birds and insects fly is so poor that further research is bound to bring new fundamental discoveries and new engineering ideas.

Notes

1. Basic notions and steady flows

1 The Deborah number was introduced by M. Reiner. All real solids contain dislocations which make them flow. Whether perfect crystals can flow under an infinitesimal shear is a delicate question, which is the subject of ongoing research. Three other states of matter – liquid, gas and plasma – flow on much shorter timescales.

2 Convection excited by a human body at room temperature is always turbulent, as can be seen in a movie in [13], section 605.

3 More details on the stability of rotating fluids can be found in section 9.4 of [2] and section 66 of [8].

4 While the equations of fluid mechanics can be generalized for a curved space, Galilean invariance takes place only in a flat space.

5 To go with a flow, using Lagrangian description, may be more difficult yet it is often more rewarding than staying on a shore. Like sport and some other activities, fluid mechanics is better doing (Lagrangian) than watching (Eulerian), according to J.-F. Pinton.

6 Actually, the Laplace equation was first derived by Euler for the velocity potential.

7 Detailed discussion of minima and maxima of irrotational flows is in [6], p. 385.

8 Presentation in section 11 of [16] is misleading in not distinguishing between momentum and quasi-momentum. It also does not include changing shape.

9 That one can use the conservation of momentum inside an elongated cylindrical surface around the solid body follows from the consideration of the momentum flux through this surface. The contribution of the pressure, $\pi \int_0^{\mathscr{R}} [p(L,r) - p(-L,r)] dr^2 = \pi \rho \int_0^{\mathscr{R}} [\dot{\phi}(-L,r) - \dot{\phi}(L,r)] dr^2 = \pi \rho \dot{u}[1 - (1 - \mathscr{R}/L)^{-1/2}]$ vanishes in the limit $L/\mathscr{R} \to \infty$. The pressure contribution does not vanish for other surfaces, see section 7.1 of [25].

10 For a potential flow the Bernoulli theorem (1.23) is a Hamilton–Jacobi equation with ϕ as action.

11 Further reading on induced mass, particle motion and quasi-momentum: [18], section 9.21 of [20] and sections 2.4–2.6. of [25].

12 The good news for busy people is that the time to burn the given number of calories is inversely proportional to the power (drag force times velocity), which is cubic in velocity – just doubling your bike speed lets you to decrease the exercise time eightfold.

13 The general statement on a zero resistance force acting on a body steadily moving in an ideal fluid sometimes is called D'Alembert paradox, even though D'Alembert established it only for a body with a head–tail symmetry.

14 The viscosity tensor η_{ijkl} is evidently symmetric under $i \leftrightarrow j$ and $k \leftrightarrow l$. Symmetry under $\{ij\} \leftrightarrow \{kl\}$ is a particular case of the Onsager relation for kinetic coefficients. If time irreversibility is broken, then the viscosity tensor in isotropic media can also have an anti-symmetric (odd) part but only in 2D. In this case, the extra stresses are $\sigma_{xx} = -\sigma_{yy} = \eta^a(\partial v_x/\partial y - \partial v_y/\partial x)$, $\sigma_{xy} = \sigma_{yx} = \eta^a(\partial v_x/\partial x - \partial v_y/\partial y)$. Velocity shear $\partial v_x/\partial y$ adds to pressure, while tangential forces are produced by compression or expansion.

15 The Navier–Stokes equation was derived by Navier in 1822, Cauchy in 1823, Poisson in 1829, Saint–Venant in 1837 and eventually by Stokes in 1845. Curiously, it bears only the names of the first and the last.

16 When can the hydrodynamic approach be used in systems with several types of collisions and respective mean free paths? For example, electrons in conductors exchange momentum among themselves but also lose it to the lattice by scattering on impurities and phonons. If the mean free path of the latter processes is smaller than that of the former, one encounters the usual Ohmic regime. However, if the electron–electron mean free path is smaller, then electrons flow like a viscous fluid.

17 "No-slip" can be seen in a movie in [13], section 605. The no-slip condition is a useful idealization in many but not all cases. Depending on the structure of the liquid and the solid and the shape of the boundary, slip can occur, which can change flow pattern and reduce drag. This rich aspect of physics, and also numerical and experimental methods used in studying this phenomenon, is described in section 19 of [29].

18 One can see liquid jets with different Reynolds numbers in section 199 of [13].

19 Movies of propulsion at low Reynolds numbers can be found in section 237 of [13]. See also [9].

20 Photographs of boundary layer separation can be found in [30] and movies in [13], sections 638–675.

21 Another familiar example of a secondary circulation due to pressure mismatch is the flow that carries the tea leaves to the center of a teacup when the tea is rotated (it was noticed by Einstein), see e.g. section 7.13 of [10] for details.

22 More details on jets can be found in sections 11, 12 and 21 of [28].

23 The shedding of eddies and resulting effects can be seen in movies in sections 210, 216, 722 and 725 of [13].

24 An elementary discussion and a simple analytic model of the vortex street can be found in section 5.7 of [2], including an amusing story told by von Kármán about the doctoral candidate (in Prandtl's laboratory) who tried in vain to polish the cylinder to make the flow non-oscillating.

25 These words and Figures 1.15 and 1.16, do not do justice to the remarkable transformations of the flow with the change of the Reynolds number; a full set of photographs can be found in [30] and movies in [13], sections 196, 216 and 659. See also Galleries at www.efluids.com/.

26 A compact presentation of the complex analysis for flows with circulation can be found in section 6.5 of [6].

27 A lively book on the interface between biology and fluid mechanics is [32].
28 Further reading on flow past a body, drag and lift: section 6.4 of [6] and section 38 of [16].

2. Unsteady flows

1 Description of numerous instabilities can be found in [8] and in Chapters 8 of [10, 22].
2 Stability analysis for pipe and plane shear flows with the account of viscosity can be found in section 28 of [16] and section 9 of [2].
3 For a brief introduction into the theory of dynamical chaos see e.g. sections 30–32 of [16], full exposure can be found in E. Ott, *Chaos in dynamical systems* (Cambridge University Press, 1992). See also Exercise 3.7.
4 Experimental data on turbulent puffs decaying and splitting in pipe flows can be seen, for instance, in [5].
5 Compact lucid presentation of the phenomenology of turbulence can be found in Sreenivasan's Chapter 7 of [22]. Detailed discussion of flux in turbulence and further references can be found in [7, 12].
6 Detailed derivation of the Kármán–Howarth relation and Kolmogorov's 4/5-law can be found in section 34 of [16] or section 6.2 of [12].
7 One can read about statistical integrals of motion of fluid particles in [11].
8 The extremum principle for the diffusion equation is the same as the Kirchhoff formulation of Ohms law with θ being the electrostatic potential and κ the conductivity.
9 While deterministic Lagrangian description of individual trajectories is inapplicable in turbulence at $Re \to \infty$, statistical description is possible and can be found in [7, 11].
10 An important application of fractal distributions produced by compressible flows is to nonequilibrium statistical physics. Hamiltonian dynamics in phase space corresponds to an incompressible flow and thus generally leads to mixing and thermalization, with coarse-graining playing the role of diffusion. By adding forcing and dissipation one deviates systems from thermal equilibrium and can create fractal distributions called Sinai–Ruelle–Bowen measures, see e.g. section IIIA4 in [11]. If one defines entropy as minus the mean logarithm of the probability density, then the sum of the Lyapunov exponents is minus the entropy production rate and thus is nonpositive – zero for incompressible and negative for compressible flows, see e.g. [7, 11].
11 Note that (65.8) in [16] is the total momentum of the fluid, while (3.54) in [10] is the momentum of the wave only.
12 Momentum and quasimomentum of a phonon are discussed in section 4.2 of [24]. For fluids, surface wave propagation is accompanied by a (Stokes) drift quadratic in wave amplitude.
13 More detailed derivation of the velocity of Riemann wave can be found in section 101 of [16].
14 Burgers equation describes also directed polymers with t being the coordinate along polymer and many other systems.
15 Despite being reducible to linear i.e. integrable, Burgers equation does not have integrals of motion except momentum. On the contrary, two other 1D universal

equations considered in the book, Korteveg–de Vries and Nonlinear Schrödinger, have infinite number of integrals of motion.

16 On experimental uses of the Döppler effect see [29].

17 Our presentation of a compressible flow past a body follows section 3.7 of [2], more details on supersonic aerodynamics can be found in Chapter 6 of [22].

18 Passing through the shock, potential flow generally acquires vorticity except when all the streamlines cross the shock at the same angle as is the case in the linear approximation, see [16], sections 112–114.

3. Dispersive waves

1 Ocean waves propagate large distances, yet they do not produce large-scale ocean currents by Stokes drift. The reason is the Coriolis force due to Earth rotation, which makes fluid to move in horizontal circles with a zero net flow.

2 Derivation of the viscous decay rate of gravity waves via stresses calculated from an ideal flow can be found in section 25 of [16].

3 For details on Stokes derivation of viscous dissipation of gravity surface waves see section 3.5 of [17].

4 Galileo was the first to determine experimentally the air density. He then related the old-known practical fact that no suction pump can lift water higher than 10 meters to the atmospheric pressure. His pupils, Torricelli and Viviani, tested this idea with a heavier liquid, such as mercury, thus creating the first barometer (in 1643). See e.g. H.S. Lipson, *The Great Experiments in Physics* (Oliver and Boyd, Edinburgh 1968).

5 On the standing wave pattern on a stream, an enjoyable reading is section 3.9 of [17] which, in particular, contains a poem by Robert Frost with a rare combination of correct physics and beautiful metaphysics. There are "books in the running brooks."

6 Moving body most effectively generates wavelengthes comparable to its size L. Those waves run with the speed $U(L) \simeq \sqrt{Lg}$ and create two maxima seen in Figure 3.4, which are analogous to the Mach-like cone with the angle $\varphi = \arcsin(U/V)$, diminishing with V. In this case U plays the role of the sound speed and the Froude number $Fr = U/V$ of the Mach number. Old manuals listed $\sqrt{gL}$ as limiting (so-called hull) speed for every ship. At hull speed the bow and stern waves interfere constructively greatly increasing wave drag (similar to sound barrier), so that old racing yachts were built long. Engines allow speedboats to exceed that speed, which is accompanied by changing the regime to the so-called planing (boat pushes water downward strong enough for the reaction force to lift the bow).

7 The details on Kelvin ship-wave pattern can be found in section 3.10 and on caustics in section 4.11 of [17].

8 The procedure for excluding nonresonant terms is a part of the theory of Poincare normal forms for Hamiltonians near fixed points and closed trajectories [4, 34].

9 More expanded exposition of two-cascade turbulence can be found in [7], detailed presentation is in [34].

10 Negative temperature can be seen in the simplest case of two groups of modes: for fixed $E = \omega_1 N_1 + \omega_2 N_2$, $N = N_1 + N_2$, the entropy $S = \ln N_1 + \ln N_2 \propto \ln(E - \omega_1 N) + \ln(\omega_2 N - E)$ gives the temperature $T^{-1} = \partial S/\partial E \propto (\omega_1 + \omega_2)N - 2E$, which is negative for sufficiently high E. For a condensate, $\omega_1 = 0$ so $T < 0$ when

$2E = 2\omega_2 N_2 > \omega_2 N$ i.e. less than half of the particles are in the condensate. Such arguments go back to Onsager who considered 2D incompressible flows in finite domains where negative temperature states correspond to clustering of vorticity into large coherent vortices.

11 Inverse cascades and persistent large-scale flow patterns exist also in rotating fluids and magnetized plasma.

12 A photo of a bore can be seen in [30].

13 Elementary discussion of hydraulic jumps can be found in section 3.9 of [2], section 2.12 of [17] or section 2.16 of [10].

14 Note that the limit of infinite depth is singular for surface waves. Indeed, $\omega^2(k^2)$ is not an analytic function for either gravity or capillary waves on a deep water.

15 Detailed presentation of the Inverse Scattering Transform can be found e.g. in Ablowitz, M. and Segur, H. *Solitons and the inverse scattering transform.*

References

[1] Ablowitz, M. and Segur, H. 2006. *Solitons and the Inverse Scattering Transform* (Philadelphia: Society for Industrial and Applied Mathematics).

[2] Acheson, D. J. 1990. *Elementary Fluid Dynamics* (Oxford: Clarendon Press).

[3] Anderson, B. D. 2008. The physics of sailing, *Physics Today*, February, 38–43.

[4] Arnold, V. 1978. *Mathematical Methods of Classical Mechanics* (New York: Springer).

[5] Avila, K., Moxey, D., de Lozar, A., Avila, M., Barkley, D. and Hof, B. 2011. The onset of turbulence in pipe flow, *Science*, **333**, 192–196.

[6] Batchelor, G. K. 1967. *An Introduction to Fluid Dynamics* (Cambridge: Cambridge University Press).

[7] Cardy, J., Falkovich, G. and Gawedzki, K. 2008. *Non-Equilibrium Statistical Mechanics and Turbulence* (Cambridge: Cambridge University Press).

[8] Chandrasekhar, S. 1961. *Hydrodynamic and Hydromagnetic Stability* (New York: Dover).

[9] Childress, S. 1981. *Mechanics of Swimming and Flying* (Cambridge: Cambridge University Press).

[10] Faber, T. E. 1995. *Fluid Dynamics for Physicists* (Cambridge: Cambridge University Press).

[11] Falkovich, G., Gawędzki, K. and Vergassola, M. 2001. Particles and fields in fluid turbulence, *Rev. Mod. Phys.*, **73**, 913–975.

[12] Frisch, U. 1995. *Turbulence: The Legacy of A. N. Kolmogorov* (Cambridge: Cambridge University Press).

[13] Homsy, G. M., Aref, H., Breuer, K. S. *et al.* 2007. *Multimedia Fluid Mechanics* (Cambridge: Cambridge University Press).

[14] Jeffreys, H. and Swirles, B. 1966. *Methods of Mathematical Physics* (Cambridge: Cambridge University Press).

[15] Lamb, H. 1932. *Hydrodynamics* (New York: Dover).

[16] Landau, L. and Lifshitz, E. 1987. *Fluid Mechanics* (Oxford: Pergamon Press).

[17] Lighthill, J. 1978. *Waves in Fluids* (Cambridge: Cambridge University Press).

[18] Lighthill, J. 1986. *Informal Introduction to Fluid Mechanics* (Cambridge: Cambridge University Press).

[19] Lipson, H. S. 1968. *Great Experiments in Physics* (Edinburgh: Oliver and Boyd).

[20] Milne-Thomson, L. M. 1960. *Theoretical Hydrodynamics* (London: MacMillan).

[21] Nazarenko, S. 2015. *Fluid Dynamics via Examples and Solutions* (Boca Raton: CRC Press).

[22] Oertel, H. ed. 2000. *Prandtl's Essentials of Fluid Mechanics* (New York: Springer).

[23] Ott, E. 1992. *Chaos in Dynamical Systems* (Cambridge: Cambridge University Press).

[24] Peierls, R. 1979. *Surprises in Theoretical Physics* (New York: Princeton University Press).

[25] Peierls, R. 1987. *More Surprises in Theoretical Physics* (New York: Princeton University Press).

[26] Purcell, E. M. 1977. Life at low Reynolds number, *Am. J. Phys.*, **45**, 3–11.

[27] Steinberg, V. 2008. Turbulence: Elastic, *Scholarpedia*, **3**(8), 5476, www.scholarpedia.org/article/Turbulence:_elastic.

[28] Tritton, D. J. 1988. *Physical Fluid Dynamics* (Oxford: Oxford University Press).

[29] Tropea, C., Yarin, A. and Foss, J. eds. 2007. *Springer Handbook of Experimental Fluid Mechanics* (Berlin: Springer).

[30] Van Dyke, M. 1982. *Album of Fluid Motion* (Stanford: Parabolic Press).

[31] Vekstein, G. E. 1992. *Physics of Continuous Media: A Collection of Problems with Solutions for Physics Students* (Bristol: Adam Hilger).

[32] Vogel, S. 1981. *Life in Moving Fluids* (Princeton, NJ: Princeton University Press).

[33] Wilczek, F. and Shapere, A. 1989. Geometry of self-propulsion at low Reynolds number, *J. Fluid Mech.*, **198**, 557–585.

[34] Zakharov, V., Lvov, V. and Falkovich, G. 1992. *Kolmogorov Spectra of Turbulence* (Berlin: Springer).

Index